Solving Engineering Problems

A Monograph covering:

- Synthesis and Analysis
- Engineering Practice Process
- Chattering Ball Problem
- Airliner Landing Problem
- Pilot Ejector Seat Problem
- Rail Truck Design Problem
- Addendum

Carl F. Zorowski
A Design Engineering Monograph

Solving Engineering Problems

Copyright 2017
All Rights Reserved

Solving Engineering Problems

The "Engineering Practice Process" is a problem-solving procedure whose objective is to promote directed mental and physical activity that uses the creative skills of synthesis and analysis to innovate effective solutions to real engineering problems.

-Z-

Solving Engineering Problems

Solving Engineering Problems

To Professors Denniston W. VerPlanck and Benjamin R. Teare of Carnegie Institute of Technology, my early career mentors and role models

-Z-

Solving Engineering Problems

Table of Contents

Chapter 1 – Synthesis and Analysis

Introduction	1
Formal Definitions	1
Comparative Processes	3
Deduction (Analysis)	3
Inverse Deduction (Analysis)	5
Induction (Science)	6
Synthesis (Design)	7

Chapter 2 – Engineering Practice Process

Introduction	9
Need Existence	9
Problem Formulation	10
Solution Concept	11
Mathematical Model and Prototype	12
Principles and Functionality	13
Compute and Collect Data	14
Checking	15
Evaluation and Generalization	16
Unacceptable Results	17
Communication of Results	18
Summary	20

Chapter 3 – Chattering Ball Problem

Introduction	23
Observed Phenomena	23
Needs Statement	25
Problem Formulation	25
Model Development	26
Mathematical Manipulation	29
Evaluating Results	33
Communication of Results	35

Chapter 4 – Airliner Landing Problem

Introduction	37
Observed Phenomena	37
Need Statement	38
Problem Formulation	38
Solution Concept	38
Model Formulation	39
Manipulation	40
Intermediate Evaluation	42
Numerical example	43
Reasonableness Check	45
Some Final Thoughts	47
Communication of Results	48

Solving Engineering Problems

Chapter 5 – Pilot Ejector Seat Problem

Introduction	53
Problem Origin	53
Needs Statement	54
Problem Definition	55
Solution Model	56
Computations for Phase 1	57
Computations for Phase 2	60
Final Kinematic Results	62
Trust Force – Phase 1	64
Trust Force – Phase 2	65
Communication of Results	66

Chapter 6 – Rail Truck Design Problem

Introduction	69
Problem Origin	69
Needs Statement	70
Problem Formulation	70
Simplifying Assumptions	71
Model Definition	71
Analysis and Manipulation	72
Numerical Calculations	74
Equal Wheel Load Redesign	75
Other Design Options	78

Another Option	79
Communication	81

Chapter 7 – Addendum
Introduction	85
Fits and Starts	85
Getting Unstuck	86
Modeling and Assumptions	88
Fundamental Principles	89

Preface

Acquiring and understanding the principles, technology and knowledge base of a specific engineering discipline is the first step in becoming an effective practitioner of the profession. This beginning is normally accomplished by completing a four year accredited undergraduate engineering degree program. The course of study typically covers subject content in defined packets of discipline knowledge. Instruction emphasizes practice in the solution of well-defined idealized problems illustrating basic principles and enhancing learning.

This idealized problem-solving environment is a far cry from daily engineering practice where real problems show little similarity to standard textbook exercises. Real problems are more complex and ill defined possessing poorly stated needs, extraneous information, conflicting goals and a myriad of other complicating factors. Solving these problems requires the skills of creativity and innovation to synthesize information and knowledge through a logical progression of tasks to arrive at meaningful results. This requires a problem solving process that receives little attention in the standard engineering curriculum.

The question raised by this dichotomy is how do engineers accomplish this transition from problem solving at the academic level to that required for successful professional practice? Currently, there is no generally accepted formal procedure to achieve this end.

At the academic level there is limited time in the curriculum and no accepted formal pedagogy to help develop individual creative and innovative skills. Recently institutions have begun to provide some instruction and practice in solving problems that simulate those experienced by practicing engineers. This represents a beginning to help young engineers make the required transition to real problem solving but is of little value to the practicing engineer.

Both the young graduate and already practicing engineer must rely on the profession and other sources to provide help in completing this transition. This assistance currently takes a variety of forms and opportunities. Industry and professional societies offer training programs, seminars, workshops and other instructional mechanisms that can help young and experienced engineers develop and improve real problem solving skills. Assistance is also available through self-study and personal improvement using available instructional media. Where available mentoring under a senior experienced engineer represents the best mechanism to make this transition.

This monograph falls into the category of self-improvement assistance. Its purpose is to serve both the young and experienced engineering practitioner.

Solving Engineering Problems

It differs from other similar publications in its unique skill's based definition of a generic problem solving process that more closely represents how engineers practice their profession. This problem solving procedure is referred to as the "Engineering Practice Process".

In this monograph the roles of synthesis and analysis as mental activities and their impact on creativity and innovation are first investigated and discussed. This is followed by the presentation of a set of generic sequential tasks that define this problem solving process. A number of real engineering problems and their solutions are included to demonstrate in detail the application of the "Engineering Practice Process" procedure. These solutions emphasize the importance and value of mathematical modeling (a dying art) and a reliance on fundamental engineering principles.

Another feature of this monograph is the use of a "stream of consciousness" style of presentation of the problem solutions. The intent is to provide a sense of the logical progression of the solver's thoughts and actions as the solution is developed and evaluated.

The mathematical manipulations included in the problem solutions should not be a deterrent to the reader. All developments are mostly algebra with some simple integration and differentiation. These analytical treatments will be of assistance in providing a better understanding of the solution process. In this era of computers and software applications operational mathematics has become a neglected friend. As engineers we should take pride in the universality of mathematics that allows us to communicate globally with professional colleagues even though we don't speak each other's native language.

The first two chapters of this monograph establish the basis for the "Engineering Practice Process" and present the eight generic sequential steps that define this problem solving procedure. The remaining chapters are devoted to examples that demonstrate its application in the solution of four real engineering problems.

Chapter1 deals with the formal definitions of synthesis and analysis as well as their meaning in the context of engineering. Also discussed is a comparison of these mental processes to identify the differences between deduction, induction and synthesis.

Chapter 2 describes in detail the eight sequential generic tasks that make up the "Engineering Practice Process" of real engineering problem solving. This includes identifying where the skills of synthesis and analysis play their role in the solution process.

Solving Engineering Problems

Chapter 3 presents a step-by-step application of the "Engineering Practice Process" problem-solving procedure to a real engineering problem. It deals with developing a useful machine calibration system based on observed physical phenomena.

Chapter 4 describes the solution of a problem that considers the feasibility of spinning up the wheels of an airplane just prior to landing to reduce tire wear.

Chapter 5 solves the problem of predicting the dynamic operational performance of a jet fighter ejection seat based on the physiological limitations of the plot.

Chapter 6 investigates the wheel to track load distribution on a proposed articulated railway truck configuration and suggests several design changes to improve its performance.

Chapter 7 deals with some important aspects of the problem solving process that are not discussed or demonstrated in the formal solution presentations in the previous chapters.

Even though the included problems reflect the discipline interest of the author their limited technical diversity should not deter from their value as a process application guide. The required engineering background has been kept at a level that

Solving Engineering Problems

will hopefully achieve that end irrespective of the discipline base of the reader.

Developing a new skill or improving an existing one takes guidance, motivation, and practice. If this monograph provides some assistance and encouragement to pursue the goal of becoming a better engineering problem solver then it will have achieved its mission.

Finally, take pride in your chosen profession and the contributions that engineering has made to the improvement of the lives of mankind and his civilization.

Carl F. Zorowski
Cary, NC
2017

Solving Engineering Problems

Chapter 1 – Synthesis and Analysis

Introduction

Solving engineering problems effectively is a highly creative endeavor. It is an intellectual process of concentrated mental and/or physical effort calling on the skills of synthesis and analysis together with an understanding of the technology and knowledge base relevant to the problem. This chapter examines the classical definitions of synthesis and analysis and their role in the context of engineering. An understanding and appreciation of these skills is critically important to providing a foundation for the engineering problem solving process

Formal Definitions

Formal definitions of words in dictionaries generally carry several explanatory statements. This helps clarify their meaning and usage. In Figure 1-1 are a few such statements for "synthesis" taken from a current Merrium-Webster Dictionary.

> Synthesis :
> 1. the composition or combination of parts or elements so as to form a whole
> 2. the combination of often diverse conceptions into a coherent whole

Figure 1-1   Synthesis - Dictionary Definition

These statements clearly imply a process of creativity. The engagement in a mental or physical activity resulting in something new that did not previously exist. This entity may be real or intellectual. It will be new and novel even though it may be a combination of elements that were previously known or understood.

In Figure 2-1 are similar statements of the definition of "analysis" taken from the same source.

> Analysis
> 1. a detailed examination of anything complex inorder to understand its nature or to determine its essentail features
> 2. separation of a whole into its component parts

Figure 1-2   Analysis – Dictionary Definition

The implication here is again a process of either physical or intellectual activity directed at taking apart and examining some existing complex entity. The purpose is to learn how it is constituted and how it functionally behaves.

These two classical definitions now need to be considered in the context of their application in the practice of engineering.

Solving Engineering Problems

Comparative Processes

Engineers deal with physical devices, processes and systems in practicing their profession. The manner in which they interact with these entities involves different intellectual activities. Knowledge of what these activities are, how they differ and what they can accomplish provides the basis for a better appreciation and understanding of the problem solving process. Consider a generic physical system and its interaction with the environment of the world around it as depicted in Figure 1-3.

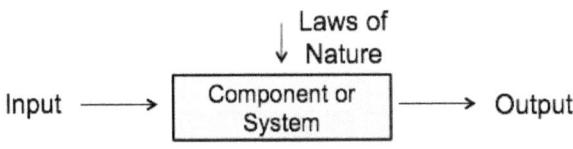

Figure 1-3   Generic System and Interactions

Subject to some specific input and the manner in which it is governed and is responsive to the laws of nature the system will generate a defined output. Depending on what is given and what is unknown about this interaction a number of intellectual processes can be recognized and identified.

Deduction (Analysis)

Begin with the example of the physical system being known, the input specified, the laws of nature governing the behavior of the system understood and the outcome being desired. As shown in Figure 1-4

Solving Engineering Problems

the intellectual process involved in determining the outcome is called deduction. In engineering this is referred to as analysis.

An example is a helical spring whose dimensions and material are given. It is assumed that the laws of nature dictate a linear behavior of the material under physical loading. If a known force is applied to the spring mathematical relations governing its behavior permit the analytical calculation of its deflection or the level of stress created. This is the intellectual process of analysis being used to determine this behavior.

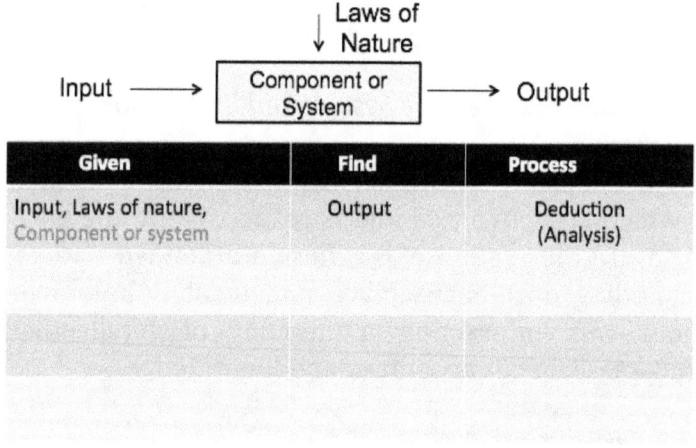

Figure 1-4   The Deductive Process of Analysis

Solving Engineering Problems

Inverse Deduction (Analysis)

Consider a second example in Figure 1-5 where the system is known, the output desired is specified and the laws of nature governing the system are understood but the input is unknown. This is in essence the inverse of the previous example and is referred to as inverse deduction. From an engineering point of view the intellectual process is still one of analysis. In the previous helical spring example this instance corresponds to a specific deflection of the spring being desired and the question is what magnitude of force will produce that outcome. This is again the process of analysis.

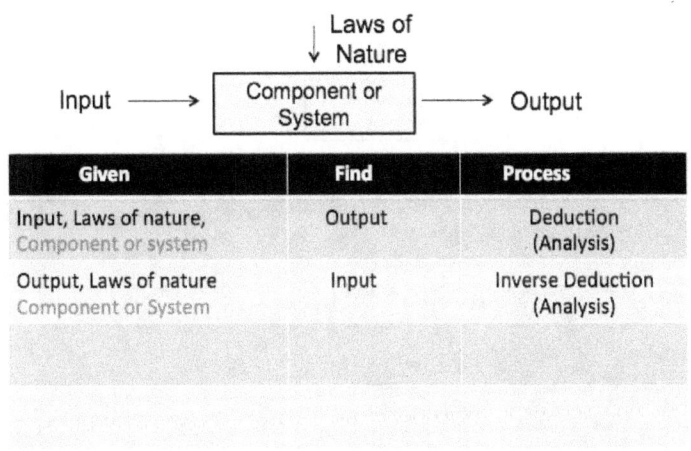

Figure 1-5   Inverse Deductive Process of Analysis

Solving Engineering Problems

Induction (Science)

A third example is when the system is known, the input is specified and the outcome is observed and measured. The unknown in this instance is the governing law of nature. The intellectual process involved is call "induction" as indicated in Figure 1-6. This is the regime of activity of scientists. The process is one of measuring experimentally the inputs and outputs of a physical system and proposing a possible theory for the behavior. The hypothesis is then tested in other instances to determine if it is a valid description of the governing behavior. If no instances are found in which the proposed theory does not work it becomes accepted as a known law of nature.

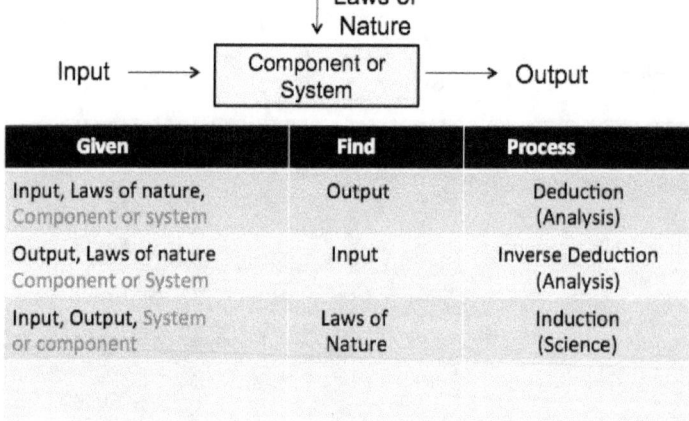

Given	Find	Process
Input, Laws of nature, Component or system	Output	Deduction (Analysis)
Output, Laws of nature Component or System	Input	Inverse Deduction (Analysis)
Input, Output, System or component	Laws of Nature	Induction (Science)

Figure 1-6   Inductive Process of Science

In the example of the helical spring this would correspond to measuring that the deflection is linearly proportional to the applied load and coming up with what in essence is Hooke's Law.

Synthesis (Design)

The final example in Figure 1-7 is when the input and desired output are both specified, the laws of nature are know and a system that satisfies these specifications is unknown. The intellectual process involved in this instance is known as synthesis. This is the act of creating something physical that didn't previously exist. In engineering this is called design.

In the example of the helical spring it is the creative activity required to specify the geometry, dimensions and material of construction such that a given applied force will produce a specified desired deflection.

Given	Find	Process
Input, Laws of nature, Component or system	Output	Deduction (Analysis)
Output, Laws of nature Component or System	Input	Inverse Deduction (Analysis)
Input, Output, System or component	Laws of Nature	Induction (Science)
Input, Output, Laws of Nature	Component or System	Synthesis (Design)

Figure 1-7   Creative Process of Synthesis (Design)

Solving Engineering Problems

It must also be recognized that the same input-output behavior could be obtained by using a different physical system such as a leaf spring or a Bellville washer. There may be other conditions or constraints that dictate the physical form the system, possibly even one not yet in common use. Choices of configurations and material become almost limitless.

A very important subset of the last example is when an existing device or system designed to satisfy prescribed input/output specifications is now required to meet some new need such as improving its function or adding some new feature. In the example of the existing helical spring or comparable compression device there might now be a requirement that the existing device possess a nonlinear spring constant to be controlled remotely. Modifying the already existing device rather than creating a totally new replacement might accomplish this just as well. The activity associated with satisfying this new need still involves the process of synthesis. It will also involve analysis as the character and nature of the performance of this modified design must be understood and evaluated.

A great deal of engineering practice deals with this type of system or product redesign effort. This represents the source of many engineering problems faced by practicing professionals. An effective engineering problem solving process must recognize and address this need. Such a process is described in the following chapter.

Chapter 2 – Engineering Practice Process

Introduction

Chapter 2 presents and describes a generic set of sequential tasks that leads to effective solutions of the kind of problems addressed at the end of the previous chapter. It will be referred to as the "Engineering Practice Process" as this type of problem represents much of what practicing engineers are faced with.

Need Existence

The progression of the Engineering Practice Process is more easily understood by examining the need and purpose of each sequential task representing this problem solving process. This will be illustrated and explained in terms of a generic flow chart the first step of which appears in Figure 2-1.

The event that must first occur to initiate the process is that a recognized technical need or requirement is recognized and can be specified. Otherwise there is no problem to be solved. The chief requirement of the need is that solving the problem it represents is considered to be of significant value to be addressed seriously. The need can arise from a variety of sources: an operational shortcoming, introduction of new technology, satisfying a regulatory issue, recognizing a business opportunity, a safety improvement, etc.

Solving Engineering Problems

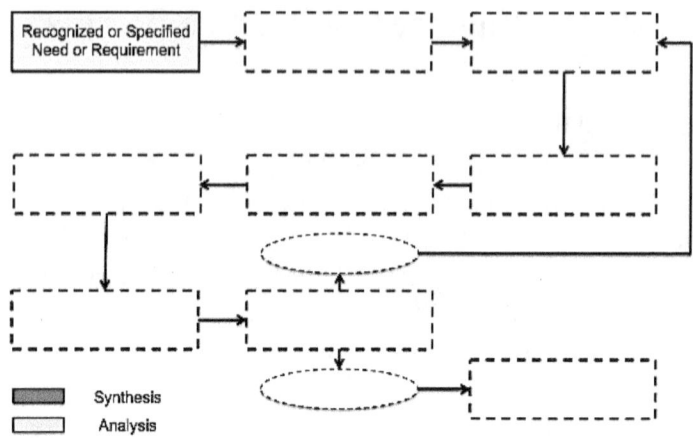

Figure 2-1   Need or Requirement Step

Problem Formulation

Following a decision to address the specified need the next step, see Figure 2-2, is to formulate a meaningful and responsive engineering problem with specific goals. This formulation should recognize the discipline subject and engineering fundamentals involved and state as operationally as possible what results are desired. For example: it is not sufficient to ask will the process get too hot if a larger heater is added to the system? Operationally, the question should be: what will be the temperature of the product if a 1000 watt heater is added to the system and run at full capacity? The very nature of the uniqueness of the problem definition to meet a specific need in measurable terms dictates the creative activity of synthesis in its formulation.

Solving Engineering Problems

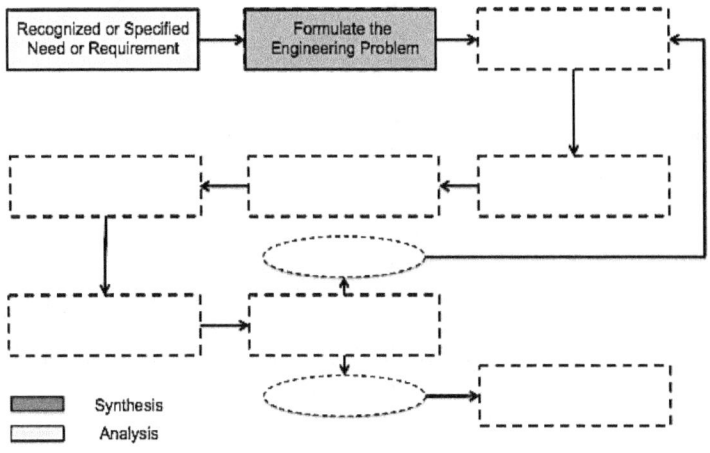

Figure 2-2   Engineering Problem Formulation

Solution Concept

Illustrated in Figure 2-3, the next step for the Engineering Practice Process is to conceptualize how a possible solution may be devised or what options are available to meet the original need. Relating to the previous step of the heater to be added to the production system might it be electrical or fed by process steam. The energy transfer mechanism might be conduction or radiation. The choice might involve physical or operational constraints and/or other pertinent considerations. Additional operational information may be needed. Creativity and innovation may be required to generate new ideas and approaches. Formulating and dealing with all possibilities and deciding on a specific solution concept proposal to proceed further involves the intellectual activities of synthesis and decision-making.

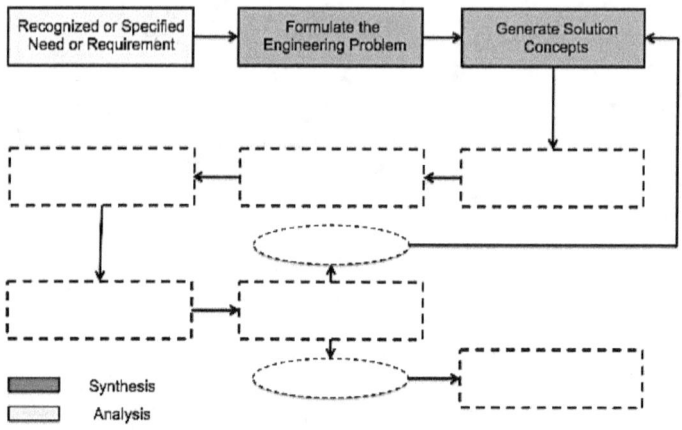

Figure 2-3 Generate Solution Concepts

Mathematical Model or Prototype

The third step in Figure 2-4 requires the creation of a mathematical model or physical prototype of the proposed solution concept that can be analyzed or tested. Only then can operational relationships be generated among the physical variables involved to produce quantifiable answers to the questions raised in the earlier definition of the problem. The uniqueness of this model as an abstracted but relevant representation of the system, process or device in question in the real world again requires the intellectual activity of synthesis.

Solving Engineering Problems

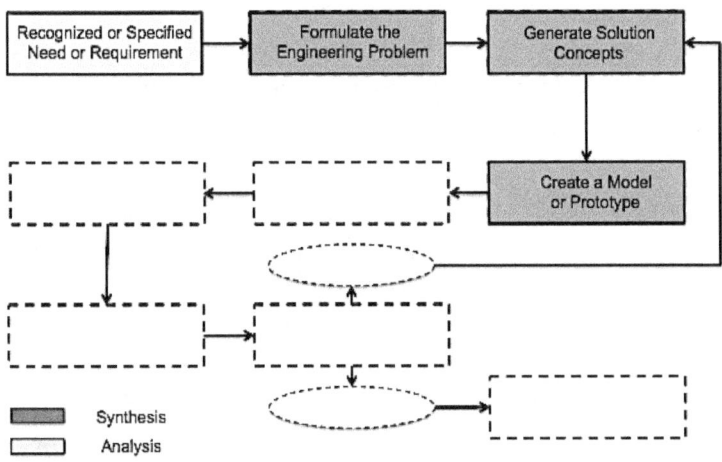

Figure 2-4   Math Model or Prototype Creation

Principles and Functionality

Once the model is established the process moves on in Figure 2-5 to providing the mathematical abstraction or experimental prototype with the appropriate technical intelligence or physical capabilities. This will permit its behavior and functionality to be analyzed. A mathematical model requires introducing symbolically the relevant principles of engineering and laws of nature that govern how the problem variables relate to one another. A problem involving mechanics may require the application of the equations of static equilibrium. In a problem involving power generation the limitations of the laws of thermodynamics might be appropriate. In a similar fashion a physical prototype must possess features of functionality that will

13

permit meaningful experimentation. This step of the total process represents a transition to the intellectual activity of analysis, as the information being added is not being newly created.

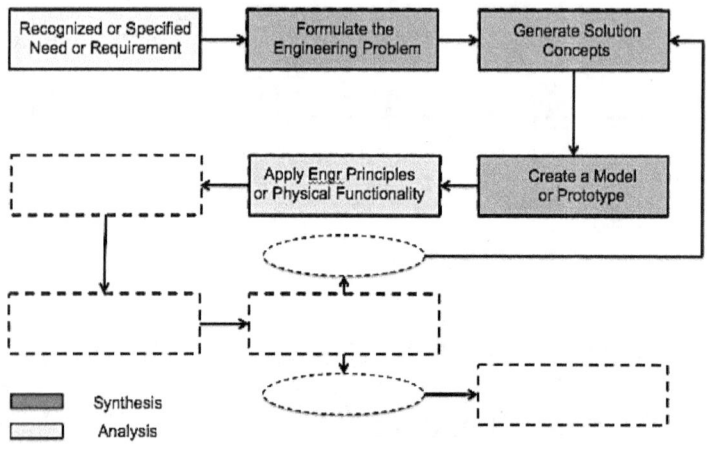

Figure 2-5 Apply Principles and Functionality

Compute and Collect Data

The model is now manipulated to extract the governing mathematical relationships between the important problem variables that permit quantitative understanding of the systems performance. At this point relevant numerical values are inserted to determine how the magnitude of the input and output parameters behave relative to one another. Data taken from tests conducted on a prototype are analyzed and compared to determine the relationship between input and output variables. These tasks are all deductive or analysis activities.

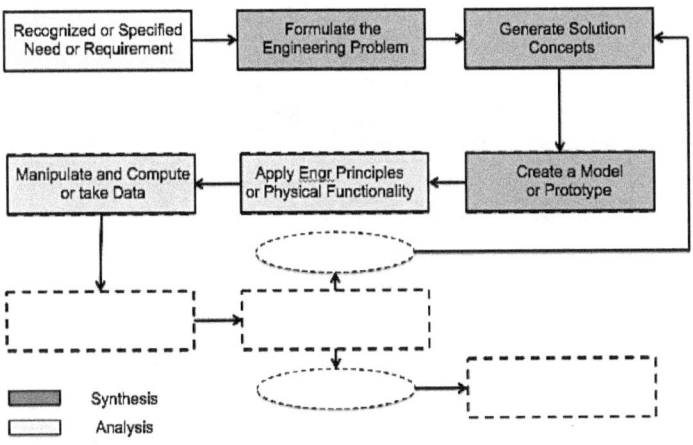

Figure 2-6   Computing and Taking Data

Checking

Although the checking task in Figure 2-7 is separated for emphasis it represents an activity that should be integrated into all aspects of the proess. It should include items like insuring unit consistency in algebraic equations, applying limiting case trend analysis to relationships between problem variables, checking algebraic manipulations, repeating numerical calculations to insure correctness, determining if variables impact performance as anticipated and reviewing the magnitude of calculated results as being physically reasonable. Again these are all deductive activities of analysis.

Solving Engineering Problems

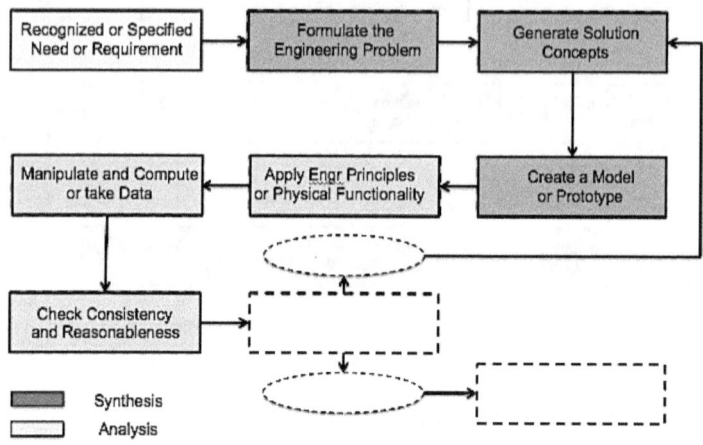

Figure 2-7 Consistency and Reasonableness Checks

Evaluation and Generalization

The evaluation and generalization task shown in Figure 2-8 is where the following questions are asked. Have the quantitative and operational questions originally posed been answered? Now that a relationship between problem variables has been developed what does it mean in terms of the general behavior of the system? Is there additional knowledge to be gained that may be of further or future use? Is it possible for generalizations to be made to give a better understanding of the behavior of the problem beyond that of the answer to the original question?

Solving Engineering Problems

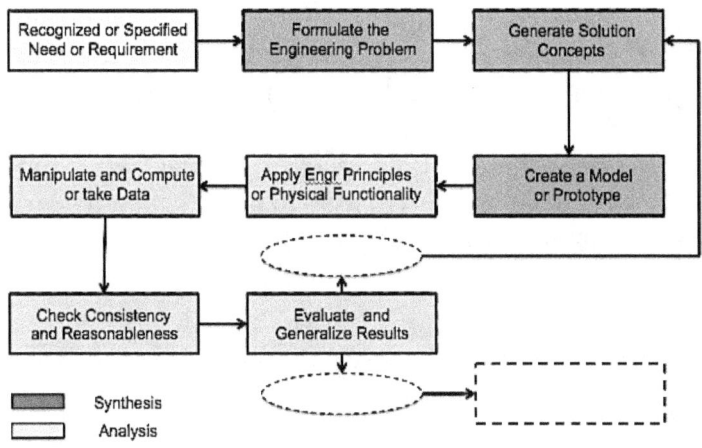

Figure 2-8   Evaluation and Generalization

Unacceptable Result

If the answer to the original problem question is unacceptable or unreasonable the problem has not been solved and the solution process must be revisited. This may require returning to the solution concept generation step, see Figure 2-9, coming up with another idea or selecting another of the several potential concepts generated originally and proceeding through the problem solving process again. This requirement of going back and repeating all or part of the process is what gives this activity its iterative character. This need for iteration can occur at any step in the process and requires returning to an early task immediately. An example would be recognizing there are insufficient equations to solve for all the unknown variables in the computation step. This might require modifying the model in the model formulation step or providing additional engineering information in the application of principles step. To rectify an error or inconsistency at any phase of the solution process dictates returning to an earlier phase to correct that difficulty before proceeding on again.

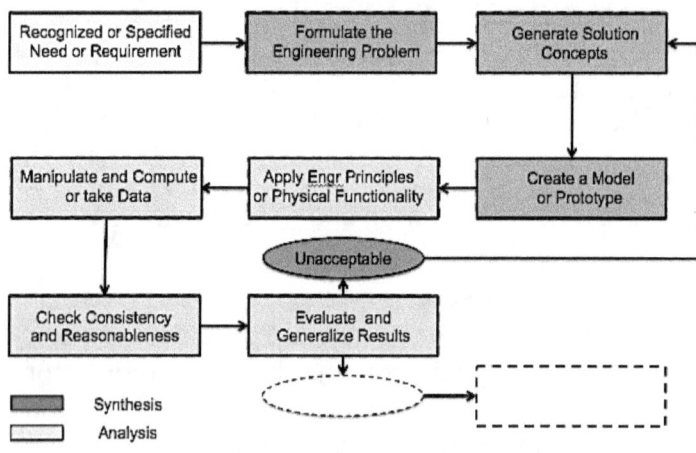

Figure 2-9   Unacceptable Answers

Communication of Results

If the answer is acceptable the last step in the process, illustrated in Figure 2-10, is to communicate the solution and any other relevant information learned in the solution process to the source of the need. This may include suggestions on how to proceed in the implementation of changes or

Solving Engineering Problems

improvements that will resolve the issue that generated the original need.

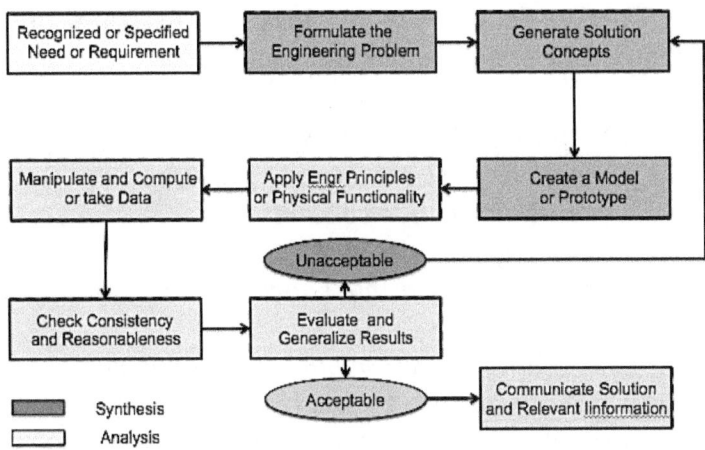

Figure 2-10   Acceptable Answer and Communication

This last task is one of the most important in the entire process and deserves special attention. An effective response or recommendation that answers the original need must be delivered in a succinct, meaningful and understandable fashion to the source of the need. It must be recognized that the recipient will not generally be interested in the intricacies and details of how the solution was created. Specific information and/or recommendations that answer the original questions asked together with the basis for their validity should represent the bulk of the communication report.

19

Summary

The eight steps of the Engineering Practice Process have been presented and discussed generically in detail as to purpose and function. In summary they are listed here once more.

Engineering Practice Process
1. *Formulating the engineering problem*
2. *Generating solution concepts*
3. *Creating a model or prototype*
4. *Applying engineering knowledge or physical functionality*
5. *Manipulating and computing or taking data*
6. *Checking consistency and reasonableness*
7. *Evaluating and generalizing results*
8. *Communicating solution and relevant information*

These tasks clearly illustrate the logical progression of the Engineering Practice Process of solving engineering problems. The application of the intellectual skills of synthesis and analysis are both necessary to the effective implementation of the process.

It now remains to demonstrate the application of this Engineering Practice Process to real problems to illustrate the importance and contribution of each step in generating an effective response to the initiating need. The rest of this monograph presents

Solving Engineering Problems

how this was accomplished with four previously solved real engineering problems.

Solving Engineering Problems

Solving Engineering Problems

Chapter 3 – Chattering Ball Problem

Introduction

The chattering ball problem deals with the need to determine if a physically observed phenomena can be used to calibrate the dynamic behavior of a mechanical testing devise. The following statement covers the specifics of the original presentation of the problem.

Observed Phenomenon

A vibration table is a machine used for subjecting small devices to a steady vibration of known amplitude and frequency. A magnetic driving force or slider crank mechanism produces the vertical motion of the table. Varying the magnetic driving force or the geometry of the sider crank changes the amplitude of the repetitive motion. A small metal ball is placed in a depression on the table. If the vertical motion is small the ball will follow the table but as the amplitude is increased a value will be reached beyond which the ball will break contact with the table and the ball will "chatter" on the plate.

Illustrated schematically in Figure 3-1 are the two vibrating table types. On the left is a magnetically driven table whose vertical amplitude is controlled by varying the power input to an electromagnet driven at a fixed frequency. On the right is a cam driven table whose amplitude is a function of the physical eccentricity of the crank mechanism.

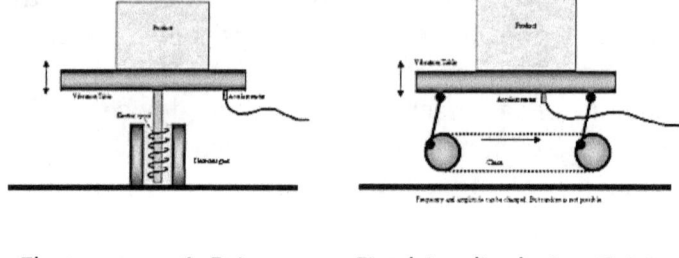

Electromagnetic Driven Table **Fixed Amplitude Cam Driven Table**

Figure 3-1 Magnetic and Mechanical Table Drives

Shown in Figure 3-2 is a magnetically driven small vertical testing table. It is observed that the

Figure 3-2 Small Magnetically Driven Table

Horizontal test plate is provided wit a number of attachment mounting holes. It is in one of these indentations that the chattering ball rests.

Needs Statement

It is suggested that the chattering behavior might be used to calibrate the amplitude of the table. The need is to investigate the phenomena and develop if possible a quantifiable analytic expression that can be used for amplitude calibration. Also any limitations to this proposed calibration technique should be investigated.

Problem Formulation and Definition

The process of synthesis is first applied to understand what is physically taking place in the observed dynamic behavior. This leads to the definition of an operational engineering problem with a specified mathematical criterion that will hopefully predict quantitatively when this phenomena takes place in terms of the system variables. Examination of this criterion is proposed through application of a specific fundamental law of engineering mechanics. This is summarized in the following formal statement of the problem definition.

> *Chattering is a consequence of the ball losing contact with the plate. This can occur if the plate moves down faster than the ball can free*

fall. When this occurs there will be no force of contact between the ball and the plate. Consider examining the equation of motion of the ball in terms of its weight and the force of contact for some specified plate motion. Apply Newton's Second Law of motion to determine the vertical equation of motion for the ball in terms of the force of contact and the prescribed motion of plate. Examine the resulting motion to determine if and when the contact force can go to zero and how this is related to the amplitude of the motion. If this results in a mathematical relationship determine numerically how it behaves and whether it is limited in some fashion.

Model Development

The next step is to create the mathematical model that will represent the ball as it undergoes vertical motion. This includes specifying the governing laws of mechanics, how the plate motion will be modeled, any physical constraints that are relevant and the criterion used to define when chatter occurs. At this point the following five items appear sufficient.

1. Ball behaves dynamically like a point mass.
2. Motion is governed by Newton's 2nd Law.
3. Only vertical motion takes place.

Solving Engineering Problems

4. Movement of the magnetically driven table can be approximated by sinusoidal harmonic motion.
5. Chatter is defined as when the force of contact between the ball and plate goes to zero.

Shown schematically on the left in Figure 3-3 is a free body diagram of the model of the system. The gray sphere represents the ball. Its weight is represented by the vector W acting vertically down, measured in lbs. force. The vector F acting upward represents the force of contact from the plate. Its units are also lbs. force. The displacement of the ball is measured by the coordinate x in the units of inches and is taken to be positive down. Since dynamic motion involves time it will be designated by "t" and measured in seconds. These defined variables should always be listed with their symbols and units with the model.

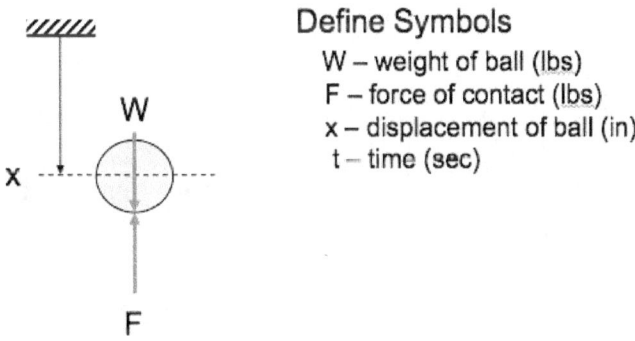

Define Symbols
W – weight of ball (lbs)
F – force of contact (lbs)
x – displacement of ball (in)
t – time (sec)

Figure 3-3 Model and Defined Variables

Solving Engineering Problems

The model now needs to be provided with intelligence as to how the weight and contact forces behave relative to one another. This is accomplished in Figure 3-4 by applying Newton's Second Law of Motion to the ball. This states that the net vertical force in the x direction is equal to the mass times the acceleration of the ball. The mass can be expressed as the weight of the ball in lbs. force divided by the acceleration due to gravity, g, in units of in./sec.² Inserting the symbolic variables into N2L gives W - F = (W/g) d²x/dt². This begins the analysis portion of the problem solution.

Apply Newton 2nd Law

$$\sum F_x = mass \times accel.$$

$$W - F = \frac{W}{g} \frac{d^2 x}{dt^2}$$

g = accel. due to gravity (in / sec²)

Figure 3-4   Applying Newton 2nd Law

The motion of the table must now be introduced into the problem. The model assumed is a simple sinusoidal harmonic motion for the magnetically driven table.

This is accomplished by representing the table displacement as x = A sin ωt where A is the amplitude in inches, omega, ω, is the angular frequency of oscillation in radians per second and t is time in seconds. The acceleration in the equation of motion is determined by differentiating this equation for the table motion twice with respect to time.

Mathematical Manipulation

Combining this expression for the second derivative with the original equation from N2L results in the final equation in Figure 3-5 involving the forces of interest and the amplitude of the table.

assume that magnetic driven motion is harmonic, i.e.

$x = A \sin \omega t$, A = amplitude of motion (in)

ω = angular frequency (rad/sec)

t = time (sec)

so that $\dfrac{d^2x}{dt^2} = -A\omega^2 \sin \omega t$ and

$$W - F = -\left(\dfrac{W}{g}\right) A\omega^2 \sin \omega t$$

Figure 3-5   Introducing Table Motion

This last equation is simplified by collecting like terms and forming a dimensionless ratio of the contact force, F, to the weight of the ball, W.

simplifying and collecting like terms

$$\dfrac{F}{W} = 1 + \dfrac{A\omega^2}{g} \sin \omega t \quad \text{unit check} \left(\dfrac{lb}{lb}\right) = \left(\dfrac{in\left(\dfrac{1}{sec^2}\right)}{in\ \dfrac{}{sec^2}}\right)$$

Figure 3-6   Simplifying into Dimensionless Terms

Solving Engineering Problems

It is appropriate at this point to conduct a unit check that is satisfied.

To obtain a better physical understanding of the behavior of the equation for F over W this ratio is plotted vertically as a function of ωt the time variable in the problem plotted horizontally in Figure 3-7. There are two parts to this relationship: the constant value of 1 and the sine curve of amplitude $A\omega^2/g$. The sine curve passes through one complete cycle at ωt equal to π and then repeats itself.

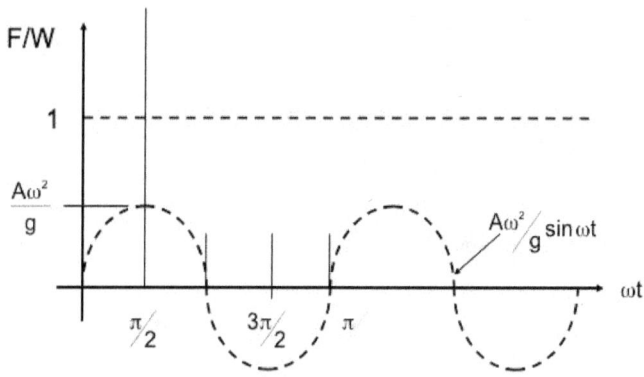

Figure 3-7   Plot of Terms in F/W Relationship

Combining the two dotted curves together results in the solid curve in Figure 3-8 representing the variation of F/W with time. It is observed that the minimum value of this ratio occurs at ωt = 3π/2. This condition will be used to apply the criterion that when F goes to zero the ball will begin to chatter.

Solving Engineering Problems

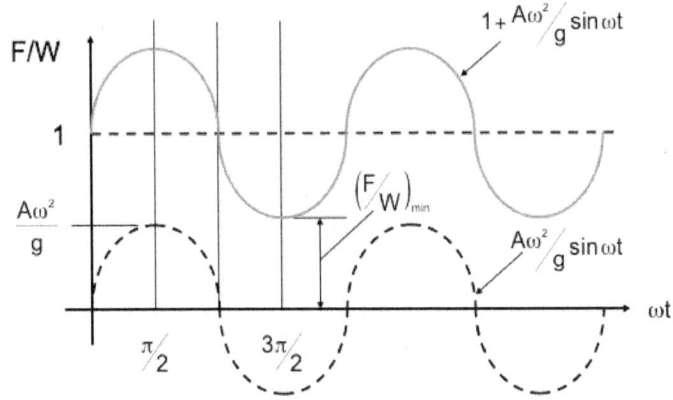

Figure 3-8 Combined Final Variation of F/W

The value of ωt equal to 3π/2 is now substituted into the general equation for F over W. The result is this minimum value of the force ratio is given by 1- A ω²/g in Figure 3-9.

$$\frac{F}{W} \text{ is a minimum at } \omega t = \frac{3\pi}{2}$$

$$\therefore \left(\frac{F}{W}\right)_{min} = 1 - \frac{A\omega^2}{g}$$

Now set $\left(\frac{F}{W}\right)_{min}$ to zero as chattering criteria, then

$$0 = 1 - \frac{A\omega^2}{g} \text{ so that}$$

$$A = \frac{g}{\omega^2} \quad \text{unit check in} = \frac{\left(\frac{in}{sec^2}\right)}{\left(\frac{1}{sec^2}\right)}$$

Figure 3-9 Amplitude / Frequency Relationship

The ratio of F/W is now set equal to zero to satisfy the chattering criterion. Solving the resulting equation for the amplitude gives A = g/ω^2 in Figure 3-9. Again units are checked for dimensional consistency.

This result indicates that at a given frequency there exist an amplitude at which chattering will occur. To be used to calibrate the table the process would be to fix the input frequency of the table and begin increasing the power to the electromagnet. As the amplitude increases chattering will eventfully occur. At the onset of chatter the amplitude can be calculated from the equation A = g/ω^2.

Numerical Computations

Numerical computations are now carried out to obtain a better understanding of how the amplitude at chatter is related to the frequency of oscillation. Driving frequencies are generally quite low for small vibrating tables. Ten to one hundred cycles per second is a reasonable range. The amplitudes at which chatter takes place for this range of frequency are listed in Figure 3-10. It is observed that this amplitude is only about a tenth of an inch at 10 cps and decreases rapidly as the frequency increases. The amplitudes become quite small at the higher frequencies. An initial conclusion is that the use of this phenomena may only be effective over a small range of frequencies.

Frequency (cps)	Ang Freq (rad/sec)	Amplitude (in)
10	63	0.098
15	94	0.043
20	126	0.024
25	157	0.016
30	188	0.011
35	220	0.008
40	251	0.006
45	283	0.005
50	314	0.004
55	345	0.003
60	377	0.003
65	408	0.002
70	440	0.002
75	471	0.002
80	502	0.002
85	534	0.001
90	565	0.001
95	597	0.001

Figure 3-10 Listed Values of Frequency vs. Amplitude

Evaluating Results

By plotting the numerical results from Figure 3-10 it is observed in Figure 3-11 that from 10 to 40 cycles per second the chattering amplitude varies much more in magnitude than it does beyond 40 cps. It is concluded that this technique of calibration would be more accurate and useful in this limited range of lower frequencies. This is an important limitation of the utility of the solution. It is a consequence of examining and evaluating numerically the realistic behavior of the generated analytic equation governing the chattering phenomena.

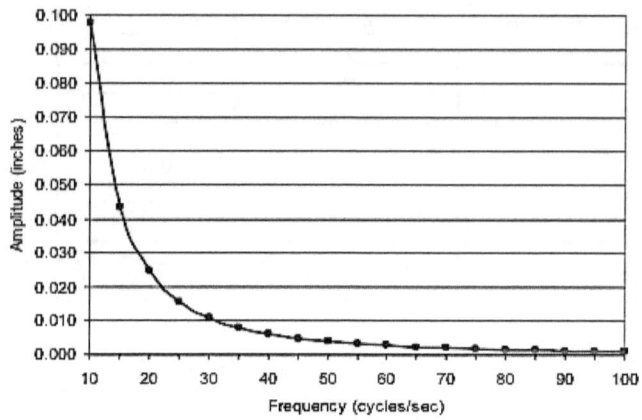

Figure 3-11 Plotted Variation of A vs. Frequency

Conclusion arrived at from the solution to this problem are listed here:

1. The amplitude at which chattering begins with a harmonic input is independent of the weight of the ball.

2. The object on the table doesn't have to be a ball providing model is applicable

3. Numerical results indicate that utility of the relationship maybe of limited use, i.e. the range of frequencies is small for which there is significant change in amplitude.

Solving Engineering Problems

Communicating Results

The remaining step is a clear and concise communication of useable results obtained from this portion of the problem solution. Following is one possible report example:

For an electromagnetic vertically driven vibration testing table the following formula can be used for determining the amplitude using the "chattering ball" phenomena:

$$A = 0.816/f^2$$

where:
A = amplitude (in)
f = frequency(cycles/sec)

Above frequencies of 40-50 cps the sensitivity of the technique is severely diminished and the accuracy of the prediction will be questionable. Use of the formula above 40 cps is <u>Not Recommended</u>.

This result <u>May Not</u> be applicable to a slider crank driven table as its movement can not be modeled as simple harmonic motion. An analytic equation for the kinematic movement of the slider crank must be introduced and the problem resolved.

Solving Engineering Problems

This last recommendation is left as an exercise for the reader.

Hints:
1. Express vertical movement x of table in terms of crank angle θ. (obtain the kinematic equation from a handbook)
2. When differentiating x to obtain acceleration remember that θ is a function of time and $d\theta/dt = \omega$, the angular frequency but ω is a constant.
3. In the manipulation phase the equations become lengthy and tedious to manage algebraically.

Chapter 4 – Airliner Landing Problem

Introduction

This problem deals with the energy dissipated in landing wheel slip when an airplane touches down on a runway.

Observed Phenomena

When an airliner touches down on landing it is observed that a "puff" of smoke emanates from each tire as it contacts the runway. It is further observed that there are black tire "skid" marks at the end of the runway where contact first takes place. These phenomena are a consequence of the tire, that is initially non-rotating, coming up to its final rotational speed corresponding to the forward velocity of the airplane.

Figure 4-1 Airliner on Landing

During this process the wheel is slipping relative to the surface of the runway. Energy is dissipated that results in excessive tire wear.

Need Statement
A proposal has been made to bring the wheel up to speed before contact with the runway to save tire wear. As a first consideration of the feasibility of this proposal it is desired to determine the energy dissipated during this initial contact between the tire and the runway as final rolling rotational speed is achieved.

Problem Formulation
When the wheel that is rotationally stationary first comes in contact with the runway slippage takes place between the wheel and the runway. Afterwards simple rolling takes place. The runway does work on the wheel in this process. This work goes into the final kinetic energy of the wheel and the energy dissipated during the slipping process. If the work done by the runway on the wheel can be calculated the energy dissipated during slippage will be this work minus the final kinetic energy of the wheel.

Solution Concept
The period of initial contact and slippage appears to occur very quickly as evidenced by the short length of the skid marks compared to the roll out distance of the plane. Hence it is reasonable to initially assume that the plane's velocity remains

constant during this time. This permits the system to be replaced by a stationary wheel coming in contact with a moving horizontal surface with a velocity equal to that of the plane.

Model Formulation

The model depicted in Figure 4-2 is a stationary wheel that can rotate about its center and a moving flat plate that is brought up in contact with it. The plate moves at a constant velocity Vo and exerts a force F on the wheel that causes it to turn. It is recognized that F varies with time but that relationship is unknown. The movement of the plate is expressed in terms of the coordinate x while the rotation of the wheel is measured by the angle θ. Both of these parameters are positive as indicated. The final angular velocity of the wheel is ω_f.

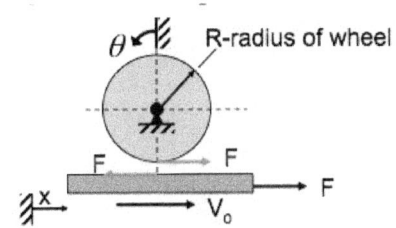

x = displacement of surface (ft)
θ = angular displacement of wheel (rad)
V_0 = surface velocity (landing speed) (ft/sec)
ω_f = final angular velocity of wheel (rad/sec)
F = F(t) = contact force (lb)

Figure 4-2   Wheel Slip Model

Manipulation

In Figure 4-3 the work done by the force F in an incremental movement of the plate can be written as dW = F dx. But dx can also be expressed as the plate velocity times an increment of time, that is V_o dt. From kinematics of rolling motion V_o is also equal to the radius of the wheel, R, times the final angular velocity of the wheel, ω_f. Combining these three relationships permits the incremental work done by the plate to be expressed as dW = F R ω_f dt. This equation would normally be integrated at this point to give the total work done but the variation of F with respect to time is not know.

Work done by force F in incremental displacement of surface of plate dx can be written as

$$dW = Fdx$$

but

$$dx = V_o dt$$

and from kinematics

$$V_o = R\omega_f$$

so that

$$dW = FR\omega_f dt, \quad (ft\ lb) = (lb)(ft)\left(\frac{rad}{sec}\right)(sec)$$

where $F = F(t)$ and is unknown

Figure 4-3 Incremental Work On Wheel

However, the dynamics of the rotational acceleration of the wheel has not yet been included. Newton's second law of motion governing rotating bodies is now applied to the wheel to introduce its rotational dynamics. The net torque producing the acceleration is simply the force, F, times the radius of the wheel, R. This is set equal to the moment of inertia of the wheel, I_w, times the acceleration expressed as $d\omega/dt$. Solving for F and substituting into the incremental work expression gives $dW = I_w \omega_f\, d\omega$ in Figure 4-4. The unknown Force as a function of time has been eliminated and the integration of dW can take place to get the total work done on the wheel.

Apply NSL $(T = I\alpha)$ to wheel to see if F an be determined or eliminated.

$$T = FR = I_w \alpha = I_w \frac{d\omega}{dt}, \quad (lb)(ft) = (ft\ lb\ sec^2)\left(\frac{rad}{sec}\right)\left(\frac{1}{sec}\right)$$

or $F = \dfrac{I_w}{R}\dfrac{d\omega}{dt}$

substitue in dW expression

$$dW = \frac{I_w}{R} R \omega_f \frac{d\omega}{dt} dt = I_w \omega_f d\omega, \quad (ft\ lb) = (ft\ lb\ sec^2)\left(\frac{rad}{sec}\right)\left(\frac{rad}{sec}\right)$$

Figure 4-4 Eliminating Unknown Force F

The incremental work is now integrated over the time of slippage to the final rotational velocity of the wheel in Figure 4-5. The result is that the total work done during slippage is equal to the moment of inertia of the wheel, I_w, multiplied by the square of the final rotational velocity of the wheel, ω_f. This total work goes into to the energy dissipated by the wheel as it comes up to speed plus its final kinetic energy, $\frac{1}{2}(I_w \omega_f^2)$. Thus the energy dissipated is seen to also be $\frac{1}{2} I_w \omega_f^2$. In other words half of the work done on the wheel is dissipated and the other half goes into its final kinetic energy of rotation.

> Now integrate over period of slippage
>
> $$\int_0^{W_{total}} dW = I_w \omega_f \int_0^{\omega_f} d\omega \quad \text{so that}$$
>
> $$W_{total} = I_w \omega_f^2, \quad (\text{ft lb}) = (\text{ft lb sec}^2)\left(\frac{\text{rad}}{\text{sec}}\right)^2$$
>
> but the total work done on the wheel is equal to the energy dissapated plus the final kinetic energy of the wheel, or
>
> $$W_{total} = I_w \omega_f^2 = E_{dis} + \frac{1}{2} I_w \omega_f^2$$
>
> $$\therefore E_{dis} = \frac{1}{2} I_w \omega_f^2$$

Figure 4-5 Energy Dissipated

Intermediate Evaluation

Some important generalizations are observed from this result:

1. The energy dissipated in wheel slip is equal in magnitude to the final rotational kinetic energy of the wheel after it comes up to rolling speed.

2. This will be true in any instance where this model represents the physical situation.

3. It was not necessary to know how the contact force varied over the period of initial contact up to final rolling speed.

4. Nothing needed to be assumed about the friction between the wheel and the runway

Solving Engineering Problems

Note that unit checks have been made while algebraically manipulating the model. This is recommended to eliminate the need to go back when an error is discovered at the end of the computations.

Numerical Example

Consider the following numerical example .to better understand the magnitude of this dissipative process A Boeing 747 has four landing wheel assemblies each consisting of four wheels. Each wheel is about six feet in diameter, weighs about 600 lbs. and has an estimated rotary inertia of 2400 lb. ft². The landing speed of the airplane is about 200 miles per hour with a landing weight of about 650,000 lbs.

First determine the rotary inertia of the wheel in the proper units by dividing its value in lb. ft.² by 32.2 ft./sec.² the acceleration due to gravity.

Hence $I_w = \dfrac{Wr^2}{g} = \dfrac{2400}{32.2} \simeq 75 \left(\text{lb ft sec}^2\right)$

with $V_o = 200 \text{ mph} \simeq 293 \left(\text{ft / sec}\right)$

then $\omega_1 = \dfrac{V_o}{R} = \dfrac{293}{3.0} \simeq 100 \left(\text{rad / sec}\right)$

so that

$$E_{dis} = \tfrac{1}{2} I_w \omega_1 = \dfrac{(75)(100)^2}{2} = 375,000 \left(\text{ft lbs}\right) \text{ per wheel}$$

Figure 4-6 Energy Dissipated per Wheel

Solving Engineering Problems

Convert the speed from 200 mph to 293 ft./sec. This makes the final rotational velocity of the wheel ω_f equal to about 100 rad./sec. The magnitude of the energy dissipated for each wheel is 375,000 ft. lbs. as shown in Figure 4-6. This is a substantial amount of energy.

For all 16 wheels the total energy dissipated is 6 million ft. lbs. The kinetic energy of the airplane at landing using the numerical data is calculated to be 900 million ft. lbs. in Figure 4-7. Since the dissipated energy at the wheels is less than one percent of the total kinetic energy of the plane the assumption that the landing velocity can be considered constant during the slip process is reasonable.

Total E_{dis} for 16 wheels

$$E_{total} = N_n \times E_{dis} = (16)(375 \times 10^3) = 6 \times 10^6 \text{ (ft lbs)}$$

Kinetic energy of plane on landing

$$KE_p = \frac{1}{2} M_w V_o^2 = \frac{1}{2} \left(\frac{650,000}{32.2} \right) (293)^2$$

$$KE_p = 900 \times 10^6 \text{ (ft lbs)}$$

∴ $E_{total} = <1\%$ of KE_p

Assumption that landing velocity V_o is constant during slip is reasonable.

Figure 4-7 Kinetic Energy of Airplane

Solving Engineering Problems

Reasonableness Check

If the time of slippage can be estimated then the length of the skid marks can also be calculated to see if they are of reasonable magnitude. This requires a model of how the force of contact varies with time to permit integration of the equation $FR = I_w \, d\omega/dt$ obtained from the application of Newton's law to the accelerating wheel.

One of the simplest models to assume is that the force F increases from zero to some maximum value F_{max} linearly with time. This can be expressed as $F(t) = F_{max}(t/t_f)$ as shown in Figure 4-8. Carrying out the integration based on this model results in t_f being equal to $2 \, I_w \, (\omega_f / F_{max} R)$.

Return to NSL $\left(T = I\alpha\right)$ applied to wheel

$$FR = I_w \frac{d\omega}{dt} \quad \text{or}$$

$$\int_0^{t_f} F(t)\,dt = \frac{I_w}{R} \int_0^{\omega_f} d\omega = \frac{I_w \omega_f}{R}$$

unfortunately F(t) is unknown so left integral can not be evaluated. As first approximation assume F(t) increase linearly from 0 to F_{max}

then $F(t) = F_{max}\left(\dfrac{t}{t_f}\right)$ so that

$$\int_0^{t_f} F(t)\,dt = \left(\frac{F_{max}}{2}\right)t_f \quad \text{and} \quad t_f = 2\,I_w \frac{\omega_f}{F_{max}R}$$

Figure 4-8 Estimate of Slippage Time

If F(t) were assumed to be F_{max} at t=0 and decrease linearly to zero at t_f exactly the same result for t_f

would be obtained, i.e. $t_f = 2\, I_w\, (\omega_f / F_{max} R)$. (This is left as an exercise for the reader to verify.)

To numerically calculate the slippage time an estimate of F_{max} is required. With little else available a reasonable approximation is that it will be of the magnitude of the coefficient of friction between the wheel and the runway and the maximum normal force that is the weight carried by each wheel. A reasonable estimate for the coefficient of friction is about 0.25 for a dry runway. Putting this all together in Figure 4-9 gives a numerical value of the slippage time to be about one third of a second. Although small it seems to be consistent with visual observation.

Now assume Coulomb friction model to approximate magnitude of F_{max}

$$F_{max} = \mu N = (\text{coef. of friction})(\text{normal force})$$

assume

$$N = \frac{W_p}{N_w} = \frac{650{,}000}{16} = 40.6 \times 10^3 \text{ lb}$$

$$\mu = 0.2 \approx 0.3$$

then $F_{max} = (0.25)(40.6 \times 10^3)\,;\ 10 \times 10^3 \text{ lb}$

resulting in $\quad t_f = 2\dfrac{I_w \omega_f}{F_{max} R} = \dfrac{(2)(100)(50)}{(10 \times 10^3)(2.75)} = 0.36 \text{ sec}$

Figure 4-9 Calculating Slippage Time Estimate

With an estimate for the time of slippage the length of the skid marks can also be approximated. Assuming the airplane's velocity remains constant during slippage the skid distance is just the velocity times the slippage time. This gives a numerical value of 105 ft. in Figure 4-10. This is a reasonable distance of observed skid mars at the ends of runways. These estimates of the magnitude of time and distance provide a sense of credibility for the analysis and results generated.

Determine skid distance as
$$D_s = V_o t_f$$
with
$$V_o = 293 \text{ (ft/sec)}$$
$$t_f = 0.36 \text{ (sec)}$$
then
$$D_s = (293)(0.36) \; ; \; 105 \text{ ft}$$
This is a reasonable distance for observed skid marks at the beginning of landing runways

Figure 4-10 Length of Skid Mark

Some Final Thoughts

The original question raised in the need statement has been answered. A formula for determining the energy dissipated as the wheel slips on the runway has been developed and the process is now much better understood. The magnitude of this dissipation has been estimated in comparison to the total kinetic energy of a large landing airliner. It appears to be only a small portion of the total amount of energy that must be dissipated in stopping a plane.

Assume that some pre-rotation device would take three minutes to bring wheel rotation up to landing speed prior to touch down. The average power required in the example cited would be 6 x10^6 ft. lbs./180 sec. or 33000 ft. lbs./sec. This is equivalent to approximately 60 horsepower. With 16 wheels on this airplane that is only about 4 horsepower per wheel. This is miniscule in comparison to he capacity of 30,000 hp. of each of its four engines. So the real question as to feasibility appears to be not one of physical capability but one of economics. Would the savings on tire wear and longer life before replacement cover the cost of design, testing, installation and maintenance of some device that would bring each wheel's rotational speed up to the plane's landing speed in a short period of time? Would the pay back period be reasonable?

Communication of Results

The question of how much energy is dissipated in wheel slip when aircraft land has been answered with a formula that permits its numerical calculation. However, a lot more has been learned about this dissipation process that should be forwarded in a report of the results. Following is an example of a how this might be done.

An analysis of a mathematical model that replicates the dynamic slipping behavior of a landing wheel at the instance of aircraft touch down resulted in the following general conclusions:

1. The energy dissipated in wheel slip is equal in magnitude to the final rotational kinetic energy of the wheel after it comes up to rolling speed.

2. This will be true in any other physical occurrence where this model is applicable.

3. It was not necessary to know how the contact force varied over the period of initial contact up to final rolling speed to arrive at this conclusion.

4. Neither was it necessary to assume anything about the friction between the wheel and the runway.

A formula for calculating the magnitude of this energy dissipation in ft. lbs. is given by:

$$E_d = 0.033\ (Wr^2)\ (V_o/R)^2$$

where:
  W – wheel weight (lbs.)
  r – radius of gyration of the wheel (ft.)
  V_o – aircraft landing Velocity (mph)
  R – radius of wheel (ft)

Solving Engineering Problems

This formula was used to estimate the magnitude of E_d for the Boeing 747 aircraft. The time of contact and the length of the skid mark on the runway were also estimated based on an assumed linear variation of the contact force and an appropriate coefficient of friction between the tire and the runway. The results of these calculations follow:

1. *The energy dissipated in wheel slip was 6×10^6 ft. lbs.*
2. *This is less than 1% of the kinetic energy of the aircraft on touch down.*
3. *The time of wheel slip was 0.36 sec.*
4. *The length of the skid mark was 105 ft.*

(Results 3 and 4 serve as checks on the solution when compared to physical observation of the phenomena.)

If it is assumed that three minutes are required for some device at each wheel to bring it to the proper rotational speed before touch down the power required by that devise would be 4 hp. This does not appear to be unreasonable for the size of the aircraft considered. It is expected that this value would be proportionally less for lighter aircraft. What is required to further evaluate this proposal is an economic analysis of

Solving Engineering Problems

whether some such device would be cost effective.

Solving Engineering Problems

Chapter 5 – Pilot Ejector Seat

Introduction

This problem deals with an analysis of the kinematics and dynamics of a jet fighter ejection seat based on the physiological limitations of the pilot.

Problem Origin

For emergency use in jet fighter aircraft it is necessary to provide a device to eject the pilot upward and out of the plane at a velocity sufficiently high to clear the tail structure. A rather dramatic example is shown in Figure 5-1.

Figure 5-1 Dramatic Example of Pilot Ejection

Solving Engineering Problems

The thrust required to achieve this ejection is provided by a specially designed rocket package attached to the back of the pilot's seat. When actuated this trust rapidly accelerates the system of pilot and seat to some specified final velocity. The process subjects the pilot to very high levels of acceleration. These must be limited to within what the human body can sustain without any permanent physical damage.

Needs Statement

A new lighter design for the rocket system is under consideration. Its operation will better match the physiological acceleration limits of the pilots. From previous designs a terminal velocity for the seat and pilot at the termination of firing should be 100 feet per second. From physiological research a pilot can withstand acceleration not exceeding 20 g for a very short period of time. In achieving this level of acceleration the rate of change of acceleration, jerk, must not exceed 150 g per second. Under these limiting conditions three questions require answers.

1. What is the vertical distance within which the ejection can be accomplished?
2. How long does it take to reach the desired final velocity?
3. What is the thrust profile required to achieve this prescribed motion?

Assume that the pilot and the ejection seat weigh about 300 lbs.

Problem Definition

The problem is to get the seat with pilot up to a velocity of 100 ft./sec. as quickly as possible. This means getting the seat up to 20 g acceleration rapidly and holding it at this value to produce a final velocity of 100 ft./sec. provided the time of duration isn't too long. To get to 20 g acceleration it is permissible to increase the acceleration at a rate of 150 g per second.

Consider the problem solution in two parts as depicted in Figure 5-2. The first phase increases the acceleration with a constant jerk of 150 g/sec to 20 g and then in a second phase 20 g acceleration is maintained to achieve the final desired velocity.

With these motions defined determine the vertical distance traveled by the seat and pilot during both phases of these motions and also the time elapsed. For simplicity assume that the plane is flying horizontal at constant velocity when ejection occurs.

If the elapsed time is sufficiently short it will not be unreasonable to assume that the effect of air drag on the pilot in the airstream can be neglected. This simplifies the analysis to only considering vertical motion.

Solving Engineering Problems

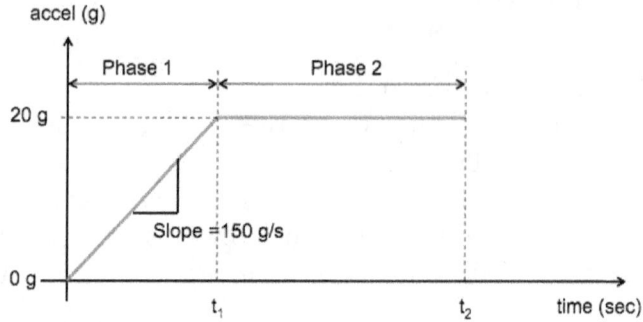

Figure 5-2 Idealized Two Phase Problem

Solution Model

Using the acceleration profile specified in Figure 5-2 the equations of kinematics that relate acceleration, velocity and distance traveled in a straight line will be solved with appropriate initial conditions to achieve 20 g in phase 1 and 100ft./per second at the end of phase 2. These equations will also permit distance traveled and time of action to be determined.

The dynamics of the pilot/seat system will be modeled using Newton's Second Law to determine the required thrust to achieve the desired acceleration profile. Again, straight vertical motion will be used based on the assumption that the total event takes place so quickly that effects of air drag can be neglected.

Computation for Phase 1

In the first phase of motion the rate at which the acceleration is increased is held at a constant jerk rate of 150 g/sec. This is expressed mathematically by setting the derivative of the acceleration with respect to time equal to a constant, A, see figure 5-3. Integrating this equation gives the acceleration equal to At plus a constant of integration. Since acceleration is the derivative of velocity with respect to time it can be integrated again to give the velocity as $V = At^2/2 + C_1t + C_2$. Finally the velocity, which is the time derivative of distance, can be integrated a third time to give the displacement $x = At^3/6 + C_1t^2/2 + C_2t + C_3$. This process has resulted in the creation of three constants of integration that must be determined from the initial conditions of the motion.

$$\frac{da}{dt} = A \text{ (const)}$$

Integrating gives accealeration (a), velocity (v) and distance (s) gives

$$a = At + C_1$$

$$v = \frac{At^2}{2} + C_1 t + C_2$$

$$x = \frac{At^3}{6} + \frac{C_1 t^2}{2} + C_2 t + C_3$$

Figure 5-3 Integration of Constant Jerk

A simplified free body diagram of the pilot and seat are shown on the left of Figure 5-4. The two forces acting on the model are the weight of the system acting vertically down and the thrust to propel the device acting upward. Displacement will be measured positive upward by the coordinate x. Applying Newton's Second Law to the system simply gives T - W = (W/g) a.

At the onset of motion when time, t, is zero the thrust is just equal to the weight of the system and the displacement, velocity and acceleration are all initially zero.

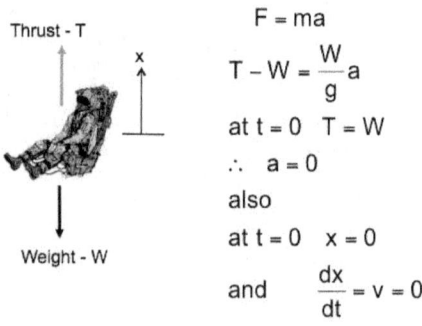

Figure 5-4 System Dynamics and I.C.'s

At the initiation of ejection the displacement x, the velocity, V, and the acceleration, a, of the seat are all zero. Substituting these initial conditions into the equations developed for acceleration, velocity and displacement results in all three constants of integration equal to zero.

Solving Engineering Problems

at $t = 0$ $s = 0, v = 0$ and $a = 0$ with $A = 150g$
therefore
$$C_1 = 0, \quad C_2 = 0, \quad C_3 = 0$$
resulting in
$$a = At = 4830\,t \quad (ft/sec^2)$$
$$v = \frac{At^2}{2} = 2415\,t^2 \quad (ft/sec)$$
$$x = \frac{At^3}{6} = 805\,t^3 \quad (ft)$$

Figure 5-5   Phase 1 Kinematic Equations

With $A = 150$ g/sec equations for the acceleration, velocity and displacement of the seat and pilot during the period in which the acceleration is increasing are presented in Figure 5-5.

From the restriction of 20g maximum acceleration at the termination of the 150 g/sec. jerk portion of the motion the time t_1 can be determined by substituting 20 g or 644 ft./sec.² into the acceleration equation and solving for the time. The result obtained in Figure 5-6 is 0.133 sec. As was anticipated this is a very short period of time. At the end of this portion of the motion the velocity attained by the seat and pilot as well as the distance it has elevated can be determined by substituting the value of t_1 into the velocity and displacement equations. The results are a velocity of 42.7 ft. per second and a height of 1.89 ft. The velocity is not yet half the value that is desired for safe ejection.

Solving Engineering Problems

First find t_1 from condition at $t = t_1$ $a = 20g = 644$ ft/sec^2
then
$$644 = 4830 t_1$$
so that
$$t_1 = \frac{644}{4830} = 0.133 \text{ sec}$$
then
$$v = 2415(0.133)^2 = 42.7 \text{ ft/sec}$$
$$x = 805(0.133)^3 = 1.89 \text{ ft}$$

Figure 5-6 Determining t_1 and V & x at t_1

Computations for Phase 2

In the second phase of the ejection the acceleration must be held constant at 20 g. Therefore the derivative of the velocity with respect to time is set equal to a constant "a". Integrating once gives the velocity as $V = at + C_1$. Integrating the velocity to get the displacement gives $x = at^2/2 + C_1 t + C_2$. Both the time t and displacement x will be measured from zero for this second phase of motion.

The two initial conditions used to determine the constants of integration will be that the displacement is zero but the velocity is 42.7 ft./sec. from the constant jerk phase at "t" equal to zero in the second phase. These calculations are carried out in Figure 5-7.

Solving Engineering Problems

$$\frac{dv}{dt} = a \text{ (const.)} = 20g = 644 \text{ ft/sec}^2$$

integrating again gives for velocity (v) and displacement (s)

$$v = at + C_1$$

$$x = \frac{at^2}{2} + C_1 t + C_2$$

but at $t = 0$, $x = 0$ and $v = v_0 = 42.7$ ft/sec

so that $\quad C_2 = 0$ and $C_1 = v_0$

Figure 5-7   Phase 2 Calculations

The time, t_2, required for the velocity to reach 100 ft./sec. is now determined from the second phase of motion kinematic equations. This is calculated to be just 0.089 sec. in Figure 5-8. With the time required for the velocity to become 100 ft. per second the distance the seat and pilot is elevated in the second phase is calculated from the displacement equation to be 6.38 ft.

First find t_2 from condition at $t = t_2$ $v = 100$ ft/sec

from

$$v = at + v_0 = 644t + 42.7$$

then

$$t_2 = \frac{100 - 42.7}{644} = 0.089 \text{ sec}$$

so that

$$x = 10gt^2 + v_0 t = 322t^2 + 42.7t$$
$$x = 322(.089)^2 + 42.7(.089) = 6.38 \text{ ft}$$

Figure 5-8 Determining t_2 and Distance x

Final Kinematic Results

The total height to which the seat and pilot are elevated in reaching 100 ft. per second is obtained by adding together the distances traveled in the two separate phases of the motion. This gives a value of 8.24 ft. in Figure 5-9. The result for the total time of ejection is similarly obtained by adding together the time of duration of each phase of the motion. The total time is 0.222 seconds. This is probably a short enough period for the pilot to withstand the rapid jerk and the 20 g acceleration with out any physiological damage. It is of interest to note that the results are independent of the weight of the seat and pilot. This is a direct consequence of the specific constraints prescribed on the kinematics of the motion, that is, a constant jerk in phase 1 and maximum acceleration in phase 2.

Total distance traveled vertically

$$x_t = x_1 + x_2$$
$$x_t = 1.89 + 6.35 = 8.24 \text{ ft}$$

Total time elapased

$$t_t = t_1 + t_2$$
$$t_t = 0.133 + 0.089 = 0.222 \text{ sec}$$

Probably short enough time that pilot can sustain 20 g.

Figure 5-9 Final Results for Distance and Time

Depicted in Figure 5-10 are graphs of how the acceleration, velocity and displacement of the seat and pilot increase with time over the two phases of the ejection event. The velocity increases parabolically in phase 1 and linearly in phase 2. The vertical distance covered increases non-linearly over the entire time of the event.

These results will be achieved only if the idealized acceleration profile of Figure 5-2 can be generated by the rocket system providing the necessary propulsive force. The design of the rocket system is a separate problem not covered in this analysis. However, the magnitude of the force required for this specified ideal performance can be determined.

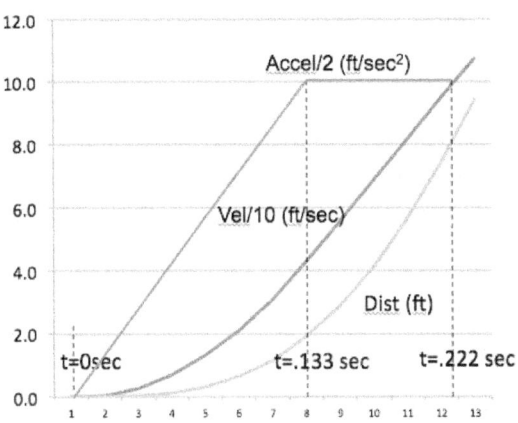

Figure 5-10 Graphical Depiction of Kinematic Results

Trust Force - Phase 1

Returning to the application of Newton's 2nd Law to the seat and pilot in Figure 5-4 the equation governing the required thrust is $T - W = (W/g)a$. The mass of the system is the weight of 300 lb. divided by the acceleration due to gravity of 32.2 ft./sec.^{2} to give 9.32 lb. sec.^{2}/ft. In the first phase of motion the acceleration is given by $a = 4830\,t$ ft./sec.^{2}

Combining these terms in Figure 5-11 gives a final equation for the thrust as $T = 300 + 4.5 \times 10^4\,t$. Note that at t equal to zero the thrust is just 300 lb. This is required to hold up the seat and pilot prior to ejection taking place. The required thrust then increases linearly with time. At end of phase 1 the thrust must be 6300 lb.

For t between $t = 0$ to $t = 0.133\,\text{sec}$

From Newton's 2nd law

$$F = ma$$

but $m = \dfrac{W}{g} = \dfrac{300\,\text{lb}_f}{32.2\,\text{ft/sec}^2} = 9.32\,\dfrac{\text{lb}_f\,\text{sec}^2}{\text{ft}}$

and $a = (4830\,t)\,\text{ft/sec}^2$ so that

$$T - 300 = 9.32\,(4830\,t)\,\text{lb}_f$$

$$T = 300 + 4.5 \times 10^4\,t$$

at $t = 0.133\,\text{sec}$

$$T = 6300\,\text{lb}_f$$

Figure 5-11   Phase 1 Thrust Calculation

Solving Engineering Problems

Thrust Force – Phase 2

During the second phase of motion the acceleration remains constant at 20 g. With constant acceleration the thrust force will also remain constant. Thus, the required thrust remains constant for this next very short period at 6300 lb.

The thrust values calculated for the two phases of motion are the magnitude of the force required to achieve the ideal kinematic motion specified. How well this thrust profile can be achieved by the rocket system used to power the ejection seat is a separate problem.

Illustrated in Figure 5-12 is how the thrust force varies over the two phases of the idealized ejection process.

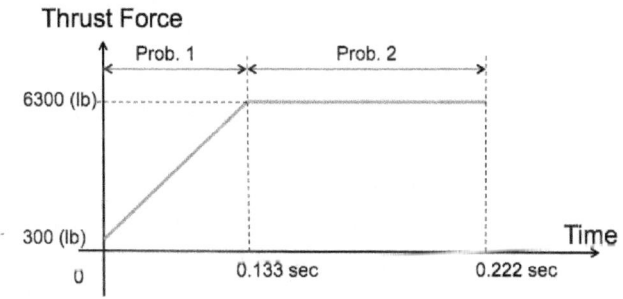

Figure 5-12 Idealized Required Thrust Profile

Of interest is that the trust begins at 300 lb. at the initiation of ejection. This of course follows from the fact that prior to the beginning of ejection this force is what supports the weight of the seat and the pilot. It is provided by a portion of the lift of the plane. Following the completion of the ejection event and rocket burn the up ward force on pilot and seat will go to zero and the system will then be subjected to the force of gravity which is simply the weight of the pilot and the chair.

Communication of Results

Following is an example report that might be used to submit the results of the analysis.

> *Researched physiological limits on pilot acceleration were used to define an idealized model to examine the kinematics of the ejection process. These included:*
> 1. *The maximum rate of acceleration increase, jerk, was taken to be 150 g/sec.*
> 2. *The maximum constant acceleration that could be sustained for a short period of time was taken to be 20g.*
>
> *This led to the acceleration profile shown.*

Solving Engineering Problems

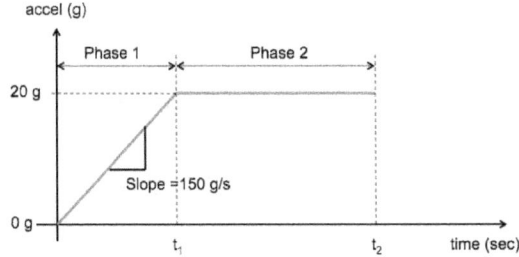

An analysis of the resulting kinematic motion resulted in the following answers to the vertical distance reached during the ejection and the time elapsed.

Phase 1
1. Time elapsed - 0.133 sec.
2. Distance traveled -1.89 ft.
3. Final velocity - 42.7 ft./sec.

Phase 2
1. Time elapsed - 0.089 sec.
2. Distance Traveled -s 6.38 ft.
3. Final Velocity -s 100ft./sec

Complete Event
1. Time elapsed w- 0.222 sec
2. Distance traveled - 8.25 ft.

A dynamic analysis of the system using the idealized acceleration profile led to the following required thrust profile.

Solving Engineering Problems

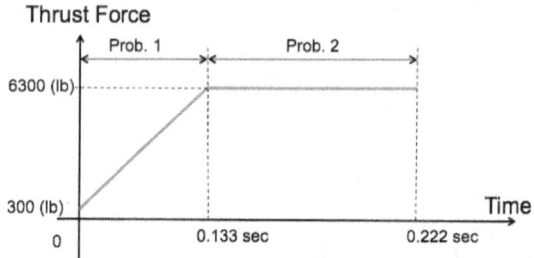

The required thrust increases linearly during the period of phase 1 to a value of 6300 lb. With the acceleration constant at 20 g during phase 2 the required thrust remains constant also at 6000 lb.

These results are based on an idealized acceleration profile dependent on the pilot's physiological limitations. Whether the results cited can actually be realized depends on how close the thrust profile can be match by the performance of the ejection rocket system.

An Interesting Extension

What happens at the end of phase 2? If the rocket thrust goes to zero the pilot will experience infinite jerk returning to –g. This is not acceptable, Assuming that -150g jerk is permissible the time required to achieve –g for the pilot will be 0.14 sec., the added distance traveled will be 18 ft. and the velocity at that point will be 142 ft./sec. It is left for the reader to validate these results.

Chapter 6 – Rail Truck Design Problem

Introduction
This problem deals with a proposed mechanical design that provides an improved kinematic feature but needs to be investigated as to what its operational loading capabilities are.

Problem Origin
A novel geometric design providing articulated flexibility has been proposed as an improvement for a Diesel locomotive wheel truck design. The geometry and dimensions of this design are presented in Figure 6-1.

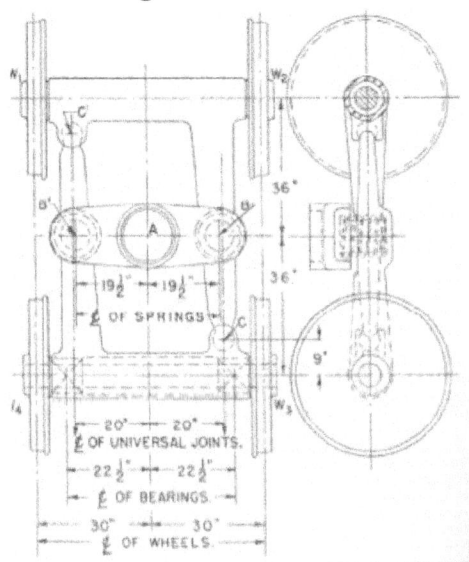

Figure 6-1 Proposed Rail Truck Design

Solving Engineering Problems

It is proposed that the ball socket joints at C and C' will permit rotation of each half of the truck about an axis through the two ball joints keeping all four wheels in closer contact with the rails even though they may be uneven.

The truck is to carry a load of 100,000 lbs. applied vertically at point A and is assumed equally divided to the two coil springs at B and B' on each half of the truck. It is important that this load be distributed equally to all four wheels irrespective of their possible vertical displacement relative to one another.

Needs Statement
Will the proposed design accomplish the desired operational goal? If it doesn't can the design be modified to provide an equal wheel load distribution?

Problem Definition
Begin by considering that the truck is on level parallel tracks loaded as given in the problem statement. If it does not provide for equal wheel loading in this ideal case it certainly won't on an uneven track. The loading carried by the wheels is a consequence of how the wheel axles are restrained by the bearings in the axle housing. If these bearing reactions are unequal so also will the wheel loadings be unequal. It seems reasonable to begin with an analysis of the static equilibrium of one of the axle

Solving Engineering Problems

housings subjected to the spring loading force, the ball joint reactions and the axle bearing reactions. This will be accomplished by applying the equations of equilibrium to these forces acting on a free body diagram of one housing unit. The analysis will be carried out in terms of general dimensions so that if the wheel reactions are not equal with the given geometry and loading the general analysis can be used to see if some simple change in the geometry will provide equal wheel loads.

Simplifying Assumptions
1. Locomotive load is distributed equally between the two springs.
2. All reactions and loadings on housing are concentrated forces and act in the vertical direction.
3. The weight of the housing will be neglected
4. The one half inch dimension between the point of action of springs and axis of housing extension will be neglected initially for simplicity.
5. Magnitude of the reaction forces at ball joints C and C' are equal as a consequence of geometric symmetry.

Model Definition
Shown in Figure 6-2 is a schematic drawing of a free body diagram of one axle-housing unit with all the loading and reaction forces indicated. Note that

Solving Engineering Problems

the ball joint reactions F_c on the short and long extensions are taken equal in magnitude but opposite in direction as dictated by symmetry.

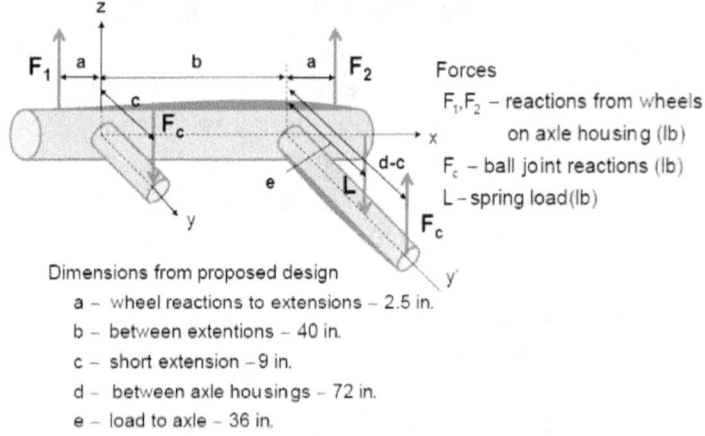

Dimensions from proposed design
 a – wheel reactions to extensions – 2.5 in.
 b – between extentions – 40 in.
 c – short extension – 9 in.
 d – between axle housings – 72 in.
 e – load to axle – 36 in.

Figure 6-2 Free Body Diagram of Single Axle Housing

This condition reduces the number of four unknown forces from four to three, two bearing reactions and a single ball joint magnitude. The dimensions for the proposed design, designated by letters, are taken from the specified geometry in Figure 6-1.

Analysis and Manipulation

In the top portion of Figure 6-3 are the three equations of equilibrium that are applicable to determine the unknown forces. They consists of vertical equilibrium of all reactions, moment equilibrium about the x axis and moment equilibrium about the y' axis. Equation (2) is first solved for Fc

Solving Engineering Problems

directly in terms of the loading, L, and housing dimensions. This expression is then substituted into equation (3) to give an equation that involves F1 and F2.

Unknown forces – F_1, F_2, F_c

Requires three equations of static equalibrium

$\sum F_z = 0 \quad F_1 + F_2 - L = 0$ (1)

$\sum M_x = 0 \quad F_c(c) - F_c(d-c) + L(e) = 0$ (2)

$\sum M_{y'} = 0 \quad F_1(a+d) - F_2(a) - F_c(b) = 0$ (3)

Solve equation (2) for F_c and substitue into equation (3) to obtain second equation relating F_1 to F_2 –

$F_c(2c - d) + L(e) = 0$

$F_c = L\left\{\dfrac{e}{(d-2c)}\right\} \qquad (lb) = (lb)\left[\dfrac{in}{in}\right]$

Figure 6-3 Equilibrium Applied to All Reactions

Next combine the equations in the first line of Figure 6-4 to give the equation on the second line.

Combine

$F_c = L\left\{\dfrac{e}{(d-2c)}\right\}$ and $F_1(a+d) - F_2(a) - F_c(b) = 0$

to give

$F_1(a+d) - F_2(a) - L\left\{\dfrac{be}{(d-2c)}\right\} = 0$

now add to equation (1) multiplied by (a) and solve for F_1

$F_1 = \dfrac{L}{(2a+b)}\left\{a + \dfrac{be}{(d-2c)}\right\}$

Figure 6-4 Solving for F_1

This equation is now added to equation (1) from Figure 6-3 multiplied by the dimension "a". The result is a solution for F_1 in terms of L and housing geometry parameters.

With F_1 determined equation (1) from Figure 6-3 can again is used to determine F_2 in Figure 6-5. Since some algebraic manipulation has been involved in these calculations it is a good ptime to do an algebra and a unit check.

$$\text{Substitue } F_1 = \frac{L}{(2a+b)}\left\{a + \frac{be}{(d-2c)}\right\} \text{ into}$$

$$F_1 + F_2 = L$$

and solve for F_2

$$F_2 = \frac{L}{(2a+b)}\left\{a + b - \frac{be}{(d-2c)}\right\} \quad (lb) = \left(\frac{lb}{in}\right)\left(in + \frac{in^2}{in}\right)$$

check

$$F_1 + F_2 = \frac{L}{(2a+b)}\left[\left\{a + \frac{be}{(d-2c)}\right\} + \left\{a + b - \frac{be}{(d-2c)}\right\}\right] = L$$

Figure 6-5 Solving for F_2 and Checking

Numerical Calculations

The housing geometry dimensions listed in Figure 6-2 are now substituted into the equations defining Fi and F2 inFigure 6 -6. The results clearly indicate that the bearing reactions on the housing are unequal. In Fact F_1 is almost double F_2. This in turn means the wheel reactions will be unequal also so the proposed articulated truck will not proform as was hoped for.

$$F_1 = \frac{L}{(2a+b)}\left\{a + \frac{be}{(d-2c)}\right\}$$

$$F_1 = \frac{50{,}000}{(2(2.5)+40)}\left\{2.5 + \frac{(40)(36)}{(72-2(9))}\right\} = 32{,}400 \text{ lb}$$

and

$$F_2 = \frac{L}{(2a+b)}\left\{a + b - \frac{be}{(d-2c)}\right\}$$

$$F_2 = \frac{50{,}000}{(2(2.5)+40)}\left\{2.5 + 40 - \frac{(40)(36)}{(72-2(9))}\right\} = 17{,}600 \text{ lb}$$

Figure 6-6 Calculated Values of F_1 and F_2

Equal Wheel Load Redesign

The value of having solved for F_1 and F_2 in general terms is that these forces can now be set equal to one another to determine if there is some set of geometric parameter values that will satisfy this condition. Setting F_1 equal to F_2 in Figure 6-7 and simplifying the result leads to the expression that $e = (d/2) - c$.

If the overall physical geometry of the axle housing is to be kept the same then referring back to Figure 6-2 the only parameter that can chnaged is the location of the spring loading "e". Keeping c = 9 in. and d = 72 in. "e" must become 27 in.

Solving Engineering Problems

$$F_1 = F_2$$

$$\frac{L}{(2a+b)}\left\{a + \frac{be}{(d-2c)}\right\} = \frac{L}{(2a+b)}\left\{a + b - \frac{be}{(d-2c)}\right\}$$

Eliminating like terms on both sides and dividing by b gives –

$$\frac{2e}{d-2c} = 1 \quad \Rightarrow \quad e = \frac{d}{2} - c \quad \text{keeping } d = 72 \text{ in}$$

then

$$e = 36 - c \quad \text{if } c = 9 \quad e = 27 \text{ in} \quad (\text{spring must be moved})$$

Figure 6-7 Determining New Spring Load Location

This relocation of the spring loading points is illustrated in Figure 6-8. This will provide for equal wheel loading without any overall change in the axle housing geometry.

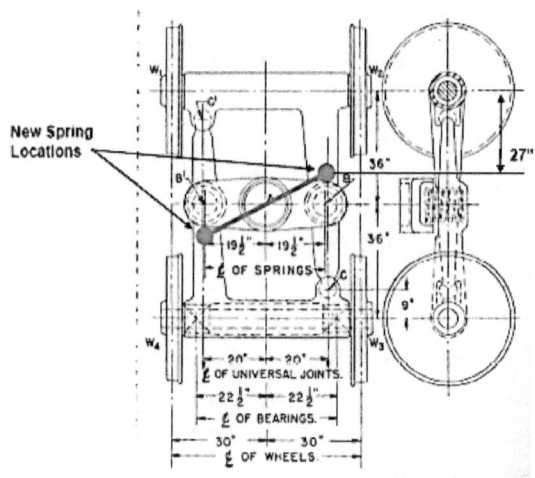

Figure 6-8 New Spring Load Locations

Solving Engineering Problems

A graphical check of this solution is demonstrated in Figure 6-9. First pass a line through the center of both ball joints. This is the axis of rotation for the truck. Next pass an axis parallel to the axis of rotation through where the left upper wheel is in contact with the track. Now consider the free body diagram of the upper left housing and sum moments of all the vertical forces about the moment axis.

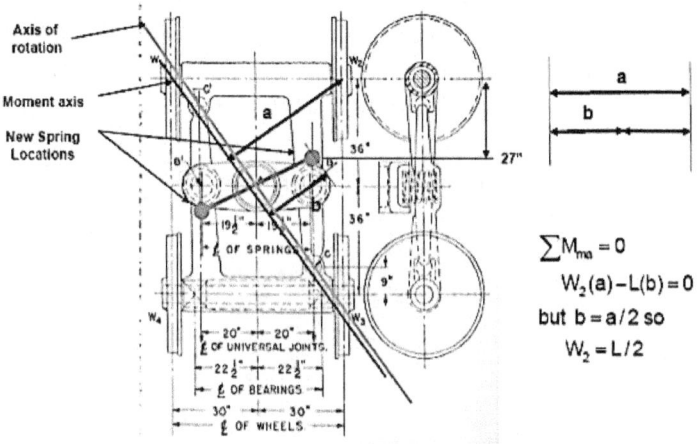

Figure 6-9   Graphic Check Of Redesign Solution

Since the left wheel force passes through the moment axis it has no moment contribution. The two ball joint forces are equal and in opposite direction at the same perpendicular distance from the moment axis. Their moment contributions cancel.

The only moment contributions left are from the spring load (50,000 lb.) and the right wheel reaction (25,000 lb.) that are opposite to each other. This requires the perpendicular distance "a" from the wheel load to be

77

twice the perpendicular distance "b" of the spring load to the moment axis to satisfy equilibrium. Measuring these two distances to scale on the drawing in Figure 6-9 shows this requirement is satisfied providing a check on the analytically derived redesign solution.

Other Redesign Options

Returning to the equation e = (d/2) – c from Figure 6-7 for equal wheel loading and keeping d =27 inches it is possible to vary "e" and "c" to satisfy this requirement. Physically "c" represents the length of the short housing extension and "e" represents the location of the spring load along the long extension from the axle centerline. Figure 6-10 lists these possible combinations.

c	e
0	36
4	32
8	28
12	24
16	20
20	16
24	12
28	8
32	4
36	0

Figure 6-10   Permissible Combinations of "c" and "e"

Solving Engineering Problems

Unfortunately, these all would require changes in the overall geometry of the axle housing extensions with the exception of the third case which is very close to the proposed overall truck geometry, The limiting case of c =0 and e = 36 inches corresponds to the ball joints being directly under the wheel axis. The other extreme of c = 36 in and e =0 puts the ball joints half way between the trucks that defeats the proposed articulation of the truck. Also this places the spring loads directly over the wheel axis.

What this variation of "c" with "e" indicates is as the short extension is increased in length with the wheel axle spacing held as it is the position of the applied spring load moves from its original proposed position of half way between the wheel axles to directly over the wheel axle for equal wheel loading.

Another Option

Presuming that a redesign change in overall truck geometry can be considered then the free body diagram geometry in Figure 6-11 illustrates another possible solution. This bent long extension keeps the spring load positions the same as the original proposed design but moves the ball joint reactions closer together along the line of the wheel axles. This new position for the short extension and the bent portion of the long extension is designated the distance "x" measured from where the bearing loads are applied at the two ends of the housing.

Solving Engineering Problems

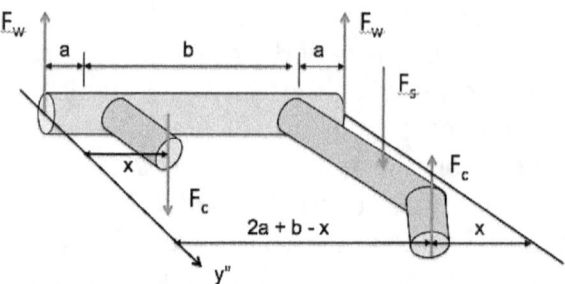

1. Keep all the dimensions the but change the location of the ball joint reactions
2. Assume that the wheel load is 25,000lb, ball joint reaction remain the same, 33,333 lbs.
3. Sum moments about y" and solve for distance x.

Figure 6-11   Another Redesign Option

The distance "x" is now determined by satisfying moment equilibrium about the y" axis in Figure 6-11.   Assuming equal wheel loads the calculation of "x" is determined to be 7.35 in.

Sum moments about y" axis

$$\sum M_{y"} = 0$$

$$F_c(x) - F_w(2a+b) + F_s(a+b) - F_c(2a+b-x) = 0$$

$$x = \frac{F_w(2a+b) + F_s(a+b) - F_c(2a+b)}{2F_c}$$

$$x = \frac{25,000(45) + 50,000(42.5) - 33,333(45)}{66,666}$$

$$x = 7.35 \text{ inches}$$

Figure 6-12   Calculation of Dimension "x"

This is not an excessive bend in the end of the long extension but does require relocating the attachment of the short housing.

Communication

An example report of the results of this analysis might be as follows:

Based on a model of the proposed truck in contact with straight parallel tracks a static equilibrium analysis was made of one half of the truck system using the free body diagram and applied forces shown in the following figure.

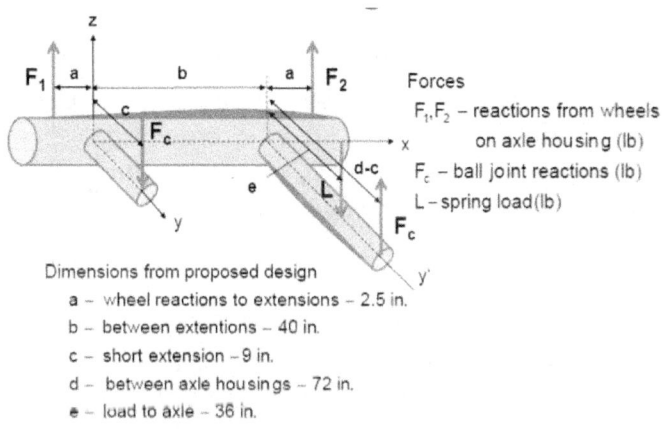

Dimensions from proposed design
- a – wheel reactions to extensions – 2.5 in.
- b – between extentions – 40 in.
- c – short extension – 9 in.
- d – between axle housings – 72 in.
- e – load to axle – 36 in.

The 50,000 lb. spring force load on the housing is applied at distance "e" from the axle axis. The ball joint forces, Fc, are considered equal in magnitude but opposite in direction from the symmetry of the entire system about the articulation axis. The analysis resulted in the two bearing forces F1 and F2 not being equal. This in turn translates to the two wheel reactions not being equal.

Thus:

THE PROPOSED TRUCK DESIGN <u>DOES NOT</u> PROVIDE EQUAL WHEEL TO TRACK LOADINGS WHETHER THE TRUCK IS ON LEVEL OR UNEVEN RAILS.

Redesign Considerations
Setting the wheel loadings equal to one another a redesign can be achieved without requiring any changes in the overall design geometry of the proposed configuration. This is accomplished by moving the location of the spring load closer to the wheel axle along the long extension.

By placing the spring contact point at 27 ins. from the wheel axis as shown in the following drawing the wheel loadings will be equal.

Solving Engineering Problems

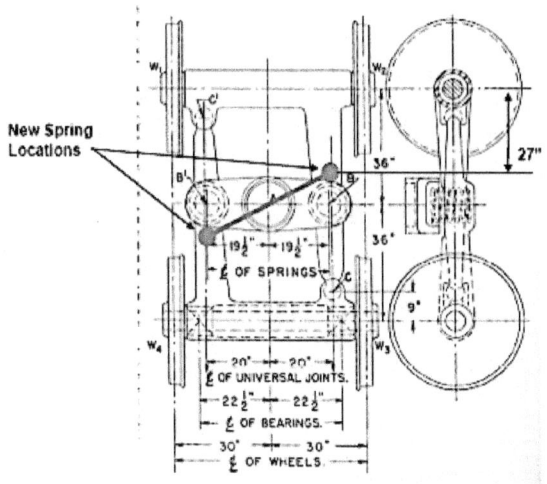

If changes in the proposed housing over- all geometry can be considered there is another way to achieve equal wheel loading without changing the original location of the spring load. Bending the end of the long extension toward the center of the truck and moving the attachment of the short extension closer to the center of the truck will accomplish this.

In the free body diagram of the housing shown in the drawing the dimension "x" defines these changes. If "x" is 7.35 inches the wheel loadings will be equal.

Solving Engineering Problems

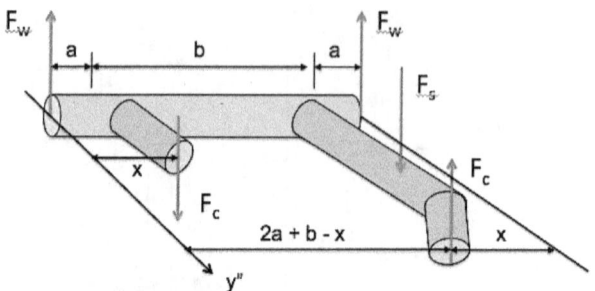

1. Keep all the dimensions the but change the location of the ball joint reactions
2. Assume that the wheel load is 25,000lb, ball joint reaction remain the same, 33,333 lbs.
3. Sum moments about y" and solve for distance x.

Other Solutions?

Can you propose any other redesigns that will achieve equal track loading on all four wheels and still retain some form of the articulation feature?

Chapter 7 - Addendum

Introduction

Experience has shown that there are some additional aspects of real problem solving not demonstrated in the examples that deserve further consideration. These issues can have an important impact on the successful application of the individual steps of the problem solving procedure.

Fits and Starts

The problem solutions in Chapters 4,5 and 6 were all presented as a very logical sequential step-by-step process from problem definition to communication. What is not demonstrated are the "fits and starts" and iterations normally experienced in real problem solving. It is easy to write up the details of how a problem was solved after an effective accepted solution is obtained. The process of getting there is never that smooth or well defined.

This does not mean the problem solving procedure is defective or ineffective. Experiencing ""fits and starts" is a normal consequence of issues like poor problem definition, unrealistic assumptions, inadequate modeling, errors in computation or incorrect evaluation. All of these can and will occur from time to time. These occurrences simply require correction by iteration back to earlier steps in the procedure. This is quite normal and should be expected rather than leading to frustration and discouragement.

Getting "Unstuck"

Another issue that often proves upsetting is that one may get to a place in a problem that taxes the solver's ability to be sufficiently creative, innovative or knowledgeable to continue making progress. The solution isn't proceeding as desired and progress seems to have hit a "brick Wall". No amount of additional repetitive effort results in any positive movement forward. The process has literally become stuck. How does one deal with this too familiar occurrence?

An obvious suggestion is to research the problem and the specific aspect of the solution that has become stuck. One never knows what information may already exist that would help. With the availability of the world's knowledge on the Internet someone may have already addressed that same problem or something similar that would be of assistance. Be sure to take advantage of this and other resources. The help you need may just be out there.

A second suggestion to is discuss the roadblock in detail with a fellow practitioner. This will require you to organize and verbalize how you see the problem and what specifically has you stuck to help the listener understand your difficulty.

Solving Engineering Problems

Questions raised or suggestions made by the listener can often help trigger a new insight for the solver. It may help clarify some aspect of the problem not previously considered or give rise to some new creative approach for overcoming the existing difficulty.

Another technique is to take advantage of what is known about how the creative process works. One popular description of the process consists of five distinct steps.

1. Preparation
2. Concentrated effort
3. Withdrawal
4. Insight
5. Follow through

The Preparation phase deals with activities like formulating the problem, gathering resources and devising a plan of attack. The Concentrated Effort step represents the implementation of the plan and expending serious effort to achieve its goal. If this is not successful in producing the desired creative output then step three comes into play. One literally walks away from the non-productive effort to some other activity. The purpose is to disengage the conscious mind from thinking further about step 2.

The subconscious mind then takes over working on what the conscious mind managed to accomplish in step 2. The result in many instances is a new creative insight that the conscious mind then recognizes, i.e. "the light bulb goes on". This is step 4 taking place. It is then up to the problem solver to follow through with this new insight and move forward. As mysterious as this process may sound it really can and does work.

Modeling and Assumptions

An effective analytic or experimental model is one that contains the important physical characteristics of the real system and can simulate the operational behavior required of the problem definition. It should be as simple as possible but sufficiently complex to include all the important physical parameters that define the behavior of the real phenomena under consideration. Achieving simplicity in the model is directly related to the solvers ability to make appropriate assumptions about the behavior of the real system. Is a variable important enough to have a first order effect on the solution? Can its influence be neglected at least initially?

As an example consider the assumption that the airplane velocity remains constant in the tire skid problem in Chapter 4 as the wheel comes up to rolling speed. This seems reasonable since the phenomenon occurs quickly. The result is the calculation of the work done by the runway on the tire is simplified. This assumption was later verified when it was determined that the time of the skid was

Solving Engineering Problems

only a third of a second. Actually, the speed of the plane would decrease even in this small time increment but the effect of this almost infinitesimal change would be of no real consequence in calculating the order of magnitude of the energy dissipated.

The problem faced by the young practitioner is how to know beforehand when an assumption about the level of impact of a problem variable can be made. Unfortunately, there are no hard and fast rules governing these decisions. It may vary from one problem to another. This is a skill that one develops with experience and knowledge from solving problems in a specific discipline area. Developing this ability is enhanced by taking the time to check out the assumption following a solution to the problem as was done in the tire skid analysis.

Fundamental Principles

Referring back to the preface, learning about a specific topic in an engineering discipline is enhanced by the practice of solving specially designed problems. In many instances the solutions involve applicable formulae derived from basic principles. This often results in developing the "bad" habit of looking for the right formula to solve a problem rather than working from fundamental principles. In real engineering problems an applicable formula most likely doesn't exist. If it did the problem would have already been solved. In the example problems included in this monograph all solutions begin with fundamental principles applied to the problem model.

Consider the jet ejection seat problem. The governing equations of kinematics were developed from the recognition that the time rate of change of distance is velocity, the time rate of change of velocity is acceleration and the time rate of change acceleration is jerk. With jerk given as a constant value reverse integration of this sequence of simple time derivatives led to equations for distance, velocity and acceleration as functions of time. With specific initial conditions identified the constants of integration where determined and final equations for phase one motion were determined. It was not necessary to try to find a specific formula for distance traveled as a function of time with rate of acceleration change being constant. Try looking for that on the internet! It can be derived, as in the problem solution, in much less time and with greater confidence.

A different problem often encountered when looking for an applicable formula is that it is often difficult to determine what simplifying or restrictive assumptions were made in its development. This can and should be of particular concern in using software that supposedly applies to your problem. The question should be raised as to what assumptions and restriction also apply there. Accuracy may also be of concern if a numerical method was used rather

Solving Engineering Problems

than a closed form analytic solution to calculate a numerical result. Also keep in mind that a computer result may give you a six figure numerical solution but more than three figures rarely has any significance in engineering practice. Including six figure numerical solutions in a report just isn't very professional.

Solving Engineering Problems

Epilog

The objective of this monograph was to provide a set of self-study improvement guidelines and example applications to assist the reader in becoming a more effective real engineering problem solver. Developing the ability to be more creative and innovative based on improved skills of synthesis and analysis occurs in a very individualistic manner. No one can tell you what the best way is to do it for yourself. A pathway can be provided but your journey to overcome the challenges and pit falls encountered must be dealt with personally. You are encouraged to pursue this journey and may you be successful in achieving your personal and professional goals.

Solving Engineering Problems

About the Author

Carl F. Zorowski is R. J. Reynolds Professor Emeritus of Mechanical and Aerospace Engineering at North Carolina State University. He attended Carnegie Mellon University receiving his doctorate in 1956. He taught at CMU until 1962 before joining North Carolina State University.

His academic carrier at NCSU included teaching, research and administration serving as Head of the Mechanical and Aerospace Engineering Department and Associate Dean for Academic Affair in the College of Engineering. He was also cofounder and Director of the Integrated Manufacturing Systems Engineering Institute, an early interdisciplinary master's practice program at NCSU.

His passion is classroom instruction, course content development and delivery methods that emphasize mathematical modeling and design performance analysis of mechanical systems. In 1993 he directed a five-year NSF funded Engineering Education Coalitions Program dedicated to revitalizing undergraduate engineering instruction that involved eight southeastern universities. Following retirement in 1997 he has offered graduate level courses in Mechanical Design Engineering in the College's distance education master's degree program.

Solving Engineering Problems

He is a fellow of the American Society of Engineering Education and the American Society of Mechanical Engineers. In 1999 he was awarded the Alexander Quarles Holladay Medal by the NCSU Board of Trustees for outstanding carrier achievement and contribution to the university.

Other monographs by the author

Design of Mechanical Power Transmissions
Design for Assembly
Design for Static Mechanical Strength
Design for Mechanical Fatigue
Design for Bending, Torsion and Buckling
Design for Deflection

Solving Engineering Problems

www.ingramcontent.com/pod-product-compliance
Lightning Source LLC
Chambersburg PA
CBHW070044210526
45170CB00012B/584